T0190459

Birkhäuser

Trends in Mathematics

Trends in Mathematics is a series devoted to the publication of volumes arising from conferences and lecture series focusing on a particular topic from any area of mathematics. Its aim is to make current developments available to the community as rapidly as possible without compromise to quality and to archive these for reference.

Proposals for volumes can be submitted using the Online Book Project Submission Form at our website www.birkhauser-science.com.

Material submitted for publication must be screened and prepared as follows:

All contributions should undergo a reviewing process similar to that carried out by journals and be checked for correct use of language which, as a rule, is English. Articles without proofs, or which do not contain any significantly new results, should be rejected. High quality survey papers, however, are welcome.

We expect the organizers to deliver manuscripts in a form that is essentially ready for direct reproduction. Any version of TEX is acceptable, but the entire collection of files must be in one particular dialect of TEX and unified according to simple instructions available from Birkhäuser.

Furthermore, in order to guarantee the timely appearance of the proceedings it is essential that the final version of the entire material be submitted no later than one year after the conference.

Eckhard Hitzer
Editor

Quaternionic Integral Transforms

A Machine-Generated Literature Overview

Editor
Eckhard Hitzer
International Christian University
Mitaka, Tokyo, Japan

ISSN 2297-0215 ISSN 2297-024X (electronic)
Trends in Mathematics
ISBN 978-3-031-28374-1 ISBN 978-3-031-28375-8 (eBook)
ISSN 2731-7323 ISSN 2731-7331 (electronic)
Research Summaries
https://doi.org/10.1007/978-3-031-28375-8

Mathematics Subject Classification: 44-xx

This book is published under the imprint Birkhäuser, www.birkhauser-science.com by the registered company Springer Nature Switzerland AG
The registered company address is: Gewerbestrasse 11, 6330 Cham, Switzerland

Foreword

This book provides a 20-year overview of work on quaternion integral transforms, including quaternion Fourier transforms. In the past, bibliographies have been created covering publications across a field of knowledge, but these are mere lists of works. A classic example is Alexander MacFarlane's *Bibliography of quaternions and allied systems of mathematics*, published in Dublin by the University Press in 1904, which provides us with a guide to the literature on quaternions from their inception in 1843 to the turn of the century around 1900. At the opposite extreme, reviews are very valuable scholarly works, providing as they do a deeper guide to a field of work and its literature, and of course, they should feature in every PhD thesis to set the scene for the work presented by the candidate. However, they are labour intensive to create, and require a deep understanding of the field, and are best suited to fairly specialized or developing topics. This book achieves something in between the extremes of the bibliography and the review, providing more than just a list of works about quaternion integral transforms, but not a deep review of the field. A key part of its value lies in the auto-summarized text that tells us more about a work than its metadata and abstract, and herein lies the novelty of what Eckhard Hitzer has done for the first time in this subject area. By taking the machine-generated summaries and categorizing them into chapters, we have a much more useful guide to the last 20 years' worth of literature on quaternion integral transforms than would be the case if it were just a bibliography like MacFarlane's. Any researcher knows that reading an abstract is often only sufficient to determine whether a work is worth accessing and reading, and that often, having obtained access to, or a copy of the work, it turns out on reading to be not as useful or relevant as had been expected. By having an automatically generated summary of the work, as here, this step can perhaps be cut out in such cases, thus leading to a saving in effort, which, after all, is what automation is supposed to achieve.

The book is, however, not just a machine generated list of summarized works. Eckhard has edited the material, by inserting some works manually to make a more complete book, and by categorizing and sorting the machine-generated results. So we have something in between the bibliography and the review, more towards the review than the bibliography. It will be useful to many researchers, I am sure.

Colchester, UK
13 December 2022

Stephen Sangwine

Introduction

Originally, I heard of the *Dimensions Auto-Summarizer* (https://www.dimensions.ai/) based on artificial intelligence from Dr. Thomas Hempfling, Editorial Director, Mathematics, Springer, at a presentation he gave at the 8th Conference on Applied Geometric Algebras in Computer Science and Engineering Geometric Algebra (AGACSE 2021), organized in Brno, Czech Republic, in September 2021. Dr. Hempfling asked if anybody would be interested to use this tool.

I later wrote him that I am interested in *quaternion Fourier transformations* (QFT) and their applications and whether the Dimensions Auto-Summarizer could be used to create a comprehensive literature survey for it. We finally agreed to turn this into a book project, and I was sent a questionnaire to start the search. I had to specify the topic in general, specific keywords, one article DOI or abstract of an article which is very relevant for the topic and relevant books.

Furthermore, as an example from another discipline, I could access part of the first book published with Springer using this machinery (Guido Visconti (ed.), Climate, Planetary and Evolutionary Sciences – A Machine-Generated Literature Overview, Springer Nature Switzerland AG, Cham, Switzerland, 2021, 346 pages) and later part of a second book (Ziheng Zhang, Ping Wang, Ji-Long Liu (eds.), CRISPR – A Machine-Generated Literature Overview, Springer Nature Singapore, 2022, 240 pages). It needs to be understood that though the Dimensions Auto-Summarizer can search all literature on the market, due to copy right limitations we can only include material published by SpringerNature. The current book appears to be the first such project in the field of mathematics.

Over time we created 26 data sets with between 1 and 955 items. We searched articles (not preprints, monographs or conference proceedings) published during the last 20 years. The Dimensions Auto-Summarizer allows the user to specify how many chapters, sections and subsections the auto summary should have and automatically selects, based on relevance score, from among all items found, clusters of publications for the sections and subsections, and aggregates a list of keywords for each section and subsection. The editor (me in this case) can generate a TOC or a full summary document and review it manually. For example, he can exclude items chosen by the Dimensions Auto-Summarizer or manually include other items, and freely regroup the TOC, add or delete chapters, sections and subsections.

Based on search results and since it is still a relatively new discipline, we decided to widen the topic to *quaternion integral transforms*. Quaternion Fourier transforms were first introduced in 1987 by Richard R. Ernst, a Swiss physical chemist, who was awarded the Nobel Prize in Chemistry in 1991 *"for his contributions to the development of the methodology of high resolution nuclear magnetic resonance (NMR) spectroscopy"* [https://www.nobelprize.org/prizes/chemistry/1991/summary/, access 23 Nov. 2022]. He introduced and applied the QFT in R. Ernst, G. Bodenhausen, and A. Wokaun. Principles of Nuclear Magnetic Resonance in One and Two Dimensions. International Series of Monographs on Chemistry. Oxford University Press, 1987. The first 25 years of fascinating QFT history (alongside the related higher dimensional Clifford Fourier transforms and wavelets) have been reviewed in F. Brackx, E. Hitzer, S.J. Sangwine, History of Quaternion and Clifford-Fourier Transforms, in: E. Hitzer, S.J. Sangwine (eds.), "Quaternion and Clifford Fourier Transforms and Wavelets", Trends in Mathematics (TIM) Vol. 27, Birkhäuser, Basel, 2013, pp. xi–xxvii. Free online text: http://link.springer.com/content/pdf/bfm%3A978-3-0348-0603-9%2F1.pdf

We furthermore decided not to rely on the Dimensions Auto-Summarizer score-based selection and automatic chapter/(sub)section grouping, but rather select all relevant items manually and also group them manually into meaningful chapters and (sub)sections. This might be quite different for a traditional main stream topic, e.g., perhaps for "Fourier transforms" in general, etc., where so many publications exist that meaningful manual grouping is too time-consuming.

We already had a mature summary with 88 items in February 2022, but delays and the ambition to be fully up to date, motivated me in September 2022 to use the free version of the Dimensions Auto-Summarizer to do a comprehensive search simply for the term "quaternion" limited to articles published in 2022, and I found over 500. Among the ones published by SpringerNature, I selected a further 22 to be included, and inserted them manually (via their DOI) into the relevant sections.

The results of the auto-generated summary are quite heterogenous, sometimes even only giving the title of a work and its DOI, or only including one or two sentences from the abstract. To make the outcome more meaningful, I have taken the freedom to simply insert the original abstract in 14 such cases (13%), always beginning with "Original Abstract".

The resulting literature summary on *quaternion integral transforms* contains five chapters. The first covers the QFT with sections on theory, uncertainty, signal processing, image processing, watermarking, saliency, and motion and video. The second chapter deals with the quaternion wavelet transform (QWT) with sections on background and theory, image processing, watermarking and forensic features, and medical images and tracking. The third chapter includes further quaternion integral transforms with sections on quaternionic curvelets, ridgelets and shearlets, on vast generalizations of the QFT known as quaternionic linear canonical transforms (QLCT), and on further transforms related to the QLCT. The fourth chapter focuses on the related topic of quaternionic moments. Finally, the fifth chapter extends to the topic of octonionic Fourier transforms, non-associative octonions being the highest (eight-)dimensional normed division algebra with quaternions as subalgebra.

The table of contents lists the titles of the publications included in each section. For each chapter, I wrote a brief chapter introduction. The Dimensions Auto-Summarizer generated lists of keywords then appear at the beginning of each chapter and (sub)section.

It is my hope that the summaries presented here may assist interested students, engineers, researchers, physicists and mathematicians to easily inform themselves comprehensively about the quickly growing field of quaternion integral transforms. Moreover, I hope that this project provides inspiration for others to use this novel artificial intelligence-based tool for undertaking further literature explorations in mathematics and natural sciences.

I finally want to express my deep gratitude to Thomas Hempfling and his colleagues Shubham Chavan, Chris Eder, Janina Krieger, Henning Schoenenberger, Tooba Shafique, Ruhina Yasmin Shaikh, Jeffrey Taub, and Vignesh Viswanathan at SpringerNature, including the technical office in India that handles the actual Dimensions Auto-Summarizer searches, for excellent introduction, instruction and collaboration on how to use this new artificial intelligence software tool.

Soli Deo Gloria

Noto Hanto, Japan
23 November 2022

Eckhard Hitzer

Contents

Quaternion Fourier Transform

Eckhard Hitzer

Chapter Introduction This chapter first focusses on the theoretical foundations of quaternion Fourier transforms. Basically, in the complex Fourier transform the imaginary complex unit i is replaced by any constant unit pure quaternion squaring to -1. Since quaternion multiplication is generically non-commutative, a kernel factor can either be placed to the right or the left of the signal to be transformed, the signal itself may be scalar or quaternion valued as well. Furthermore, two kernel factors can be used, one on the left and one on the right, yielding a two-sided QFT. Due to quaternion non-commutativity it is not trivial to derive the properties of QFTs. A recent comprehensive treatment is given in E. Hitzer, *Quaternion and Clifford Fourier Transforms*, Chapman and Hall/CRC, London, 1st edition (September 22, 2021). The one-sided QFT had been implemented for quite a while in the Quaternion Toolbox for Matlab (QTFM) by S. Sangwine, and most recently he has expanded this in version 3.4 to include the two-sided QFT as well (https://sourceforge.net/projects/qtfm/).

The focus of Sect. 1.1 is on general theory of the QFT, while Sect. 1.2 surveys a set of special theorems for the QFT.

Already for the conventional complex Fourier transform the study of position and frequency uncertainties is of great practical importance, e.g., in the field of quantum mechanics, and has proven mathematically very fruitful. Most results can be generalized to the QFT, even though proofs tend to become more elaborate, and the order of expressions has to be carefully preserved. This is the scope of Sect. 2.

A primary application of Fourier transforms is signal processing. The QFT can be applied to analyze various kinds of real, complex or quaternionic signals, in telecommunications, and speech processing, as shown in Sect. 3.

E. Hitzer (✉)
International Christian University, Mitaka, Tokyo, Japan
e-mail: hitzer@icu.ac.jp

E. Hitzer (ed.), *Quaternionic Integral Transforms*, Trends in Mathematics,
https://doi.org/10.1007/978-3-031-28375-8_1

Because colors can be very meaningfully encoded with three-dimensional pure quaternions *i, j, k*, assigning, e.g., the red, green and blue values to these axis, holistic color image processing has become a preferred application in fields like color edge detection, color image denoising, orientation estimation, feature detection, color texture, varying illumination, and even multiband satellite image analysis, as explained in Sect. 4.

The suitability of quaternions for color image processing also extends to new quaternionic color image watermarking schemes, surveyed in Sect. 5.

Section 6 treats the application of quaternion Fourier transforms to saliency detection models, with stereoscopic 3D images, for eye fixation prediction and spatiotemporal visual saliency.

Finally, Sect. 7 presents visual tracking in combination with neural networks and the creation of analytic (2D + t) video signals.

Machine Generated Keywords theorem, QFT, principle, fouri, uncertainty principle, uncertainty, quaternion fouri, two-side, two-side quaternion, saliency, fouri transform, model, spectral, transform, signal

1 Theory

Machine Generated Keywords theorem, fouri, convolution, Paley-Wiener, fouri transform, two-side quaternion, QFT, quaternion signal, quaternion fouri, operator, function, framework, two-side, Paley-Wiener theorem, spectral

1.1 General Theory

Machine Generated Keywords convolution, quaternion signal, QFT, general, transformation, two-side quaternion, fouri, spectral, operator, two-side, term, behavior, full, fouri transformation, section

1.1.1 Quaternion Fourier Transform on Quaternion Fields and Generalizations

DOI: https://doi.org/10.1007/s00006-007-0037-8

Abstract-Summary
We treat the quaternionic Fourier transform (QFT) applied to quaternion fields and investigate QFT properties useful for applications.

We relate the QFT computation for quaternion fields to the QFT of real signals.

We research the general linear (GL) transformation behavior of the QFT with matrices, Clifford geometric algebra and with examples.

Acknowledgement
A machine generated summary based on the work of Hitzer, Eckhard M. S.
2007 in Advances in Applied Clifford Algebras

1.1.2 General Two-Sided Quaternion Fourier Transform, Convolution and Mustard Convolution

DOI: https://doi.org/10.1007/s00006-016-0684-8

Original Abstract In this paper we use the general two-sided quaternion Fourier transform (QFT), and relate the classical convolution of quaternion-valued signals over R^2 with the Mustard convolution. A Mustard convolution can be expressed in the spectral domain as the point wise product of the QFTs of the factor functions. In full generality do we express the classical convolution of quaternion signals in terms of finite linear combinations of Mustard convolutions, and vice versa the Mustard convolution of quaternion signals in terms of finite linear combinations of classical convolutions.

Acknowledgement
A machine generated summary based on the work of Hitzer, Eckhard
2016 in Advances in Applied Clifford Algebras

1.1.3 Quaternionic Wiener–Khinchine Theorems and Spectral Representation of Convolution with Steerable Two-Sided Quaternion Fourier Transform

DOI: https://doi.org/10.1007/s00006-016-0744-0

Original Abstract In this paper we use the general two-sided quaternion Fourier transform (QFT) in order to derive Wiener–Khinchine theorems for the cross-correlation and for the auto-correlation of quaternion signals. Furthermore, we show how to derive a new four term spectral representation for the convolution of quaternion signals.

Acknowledgement
A machine generated summary based on the work of Hitzer, Eckhard
2016 in Advances in Applied Clifford Algebras

1.1.4 The Quaternion Domain Fourier Transform and Its Properties

DOI: https://doi.org/10.1007/s00006-015-0620-3

Abstract-Summary
It seems natural to define a quaternion Fourier transform for quaternion valued signals over quaternion domains.

This quaternion domain Fourier transform (QDFT) transforms quaternion valued signals (for example electromagnetic scalar-vector potentials, color data, space-time data, etc.) defined over a quaternion domain (space-time or other 4D domains) from a quaternion position space to a quaternion frequency space.

Acknowledgement
A machine generated summary based on the work of Hitzer, Eckhard 2015 in Advances in Applied Clifford Algebras

1.1.5 Multilinear Localization Operators Associated to Quaternion Fourier Transforms

DOI: https://doi.org/10.1007/978-3-319-47512-7_5

Abstract-Summary
We study the multilinear localization operator $L_{\varphi,\psi}{}^{F}$ f associated to the quaternion Fourier transform (QFT).

Introduction
Inspired by [1], we define the localization operator associated to the quaternion Fourier transform, and study the boundedness and compactness of it.

Although the commutative law is not always right for general quaternions, such as $e^{i} \cdot e^{j} \neq e^{i+j}$, the following identity related to the commutative law is still right: $e^{\mu a} \cdot e^{\mu b} = e^{\mu(a+b)} = e^{\mu b} \cdot e^{\mu a}$, where a, b are reals, μ is a unit pure quaternion, $e^{a\mu} = \cos a + \mu \sin a$.

We also have the following inequality: If μ is a unit pure quaternion, then in order to study Wigner transforms and localization operators, we first recall some properties about the quaternion Fourier transform.

The difference with the traditional Fourier transform is that the imaginary element number i in the exponent is replaced by a unit pure quaternion μ.

The parallel and perpendicular decomposition holds for any unit pure quaternion μ.

Quaternionic Multilinear Localization Operators
Each localization operator $L_{\varphi,\psi}{}^{F}$ f depends on a symbol function F, an analysis window functions φ, and a synthesis window ψ.

With fixed F, it can be viewed as a multilinear operator $L_{\cdot,\cdot}{}^{F}\cdot$.

- ☐ Like [2], we introduce the following results, and we will use them to prove the compactness of multilinear localization.
- ☐ If we fix F, then ψ, $L_{\varphi,\psi}{}^{F}$ f (t) can be viewed as an operator acting on F and φ.

Acknowledgement
A machine generated summary based on the work of Ma, Guangsheng; Zhao, Jiman
2017 in Trends in Mathematics

1.1.6 Quaternion Fourier Transform and Generalized Lipschitz Classes

DOI: https://doi.org/10.1007/s00006-020-01098-0

Introduction
His results were generalized by many authors from the viewpoint of the generalization of condition on the majorant of the modulus of continuity.

Móricz [3, 4–5] studied the continuity and smoothness properties of a function f with absolutely convergent Fourier series.

Volosivets published two papers, see [6, 7], in which he generalized all previous results.

Our aim is to generalize those results for the quaternion Fourier transform.

Preliminaries
We briefly collect some definitions and facts relevant for the quaternion Fourier transform.

For more details we refer the reader to [8, 9–10, 11, 12].

Conclusion
A characterization of the generalized Lipschitz spaces in terms of decay of the quaternion Fourier transform is given.

These types of results are, nowadays, known as Boas-type results since it was R.P. Boas who, in 1967, proved the first characterization of this type.

Acknowledgement
A machine generated summary based on the work of Loualid, El Mehdi; Elgargati, Abdelghani; Daher, Radouan
2021 in Advances in Applied Clifford Algebras

1.1.7 The Quaternion Fourier and Wavelet Transforms on Spaces of Functions and Distributions

DOI: https://doi.org/10.1007/s40687-020-00209-4

Abstract-Summary
The continuous quaternion wavelet transform of periodic functions is also defined and its quaternion Fourier representation form is established.

Introduction
The inverse right-sided QFT of f is given. For various properties of the quaternion Fourier transformation, we may refer to [10, 13, 14, 15].

E. S. M. Hitzer has derived a directional uncertainty principle for quaternion valued functions subject to the quaternion Fourier transformation [16].

The quaternion Fourier transform is also used in image and signal processing [17, 18].

[Section 2]
We have studied the right-sided QFT of test functions.

This shows the continuity of the right-sided QFT.

Continuous Quaternion Wavelet Transform of Periodic Functions
The continuous quaternion wavelet transform (CQWT) is a generalization of the classical continuous wavelet transform.

For various properties of the CQWT, we may refer to [19, 20, 21, 13, 22].

We have studied the continuous quaternion wavelet transform of periodic functions.

Acknowledgement
A machine generated summary based on the work of Lhamu, Drema; Singh, Sunil Kumar

2020 in Research in the Mathematical Sciences

1.2 Special Theorems

Machine Generated Keywords theorem, Paley-Wiener, framework, Paley-Wiener theorem, fouri transform, chapter, fouri, quaternion fouri, paper establish, quaternion-valued function, quaternion-valued, theorem quaternion, sense, transform, class

1.2.1 Titchmarsh's Theorem and Some Remarks Concerning the Right-Sided Quaternion Fourier Transform

DOI: https://doi.org/10.1007/s40590-019-00274-y

Abstract-Summary
Abilov and others (Comput Math Math Phys 48:2146, 2008) proved two useful estimates for the Fourier transform in the space of square integral multivariable functions on certain classes of functions characterized by the generalized continuity modulus, and these estimates are proved by Abilov for only two variables, using a translation operator.

The purpose of this paper is to study these estimates for Quaternion Fourier transforms, also the functions satisfy Lipschitz conditions of certain orders.

Introduction
The quaternion Fourier transform (QFT) is an extension of the classical two-dimensional Fourier transform (FT) in the framework of quaternion algebra.

With the recent popularity of quaternion and quaternion Fourier transforms (QFTs) technics in physics and engineering applications, there tends to be an inordinate degree of interest placed on the properties of QFTs.

The quaternionic Fourier transform (QFT) plays a vital role in the representation of signals.

Understanding the QFT paves the way for understanding other integral transform, such as the quaternion fractional Fourier transform (QFRFT) [23, 24, 25, 26, 27], by quaternion linear canonical transform (QLCT) [28] and the quaternion Wigner–Ville distribution [29].

It has become popular to generalize the Fourier transform (FT) from real and complex numbers [30] to quaternion algebra.

To describe our results, we first need to introduce some facts about harmonic analysis related to the (right-sided) quaternion Fourier transform (QFT).

Acknowledgement
A machine generated summary based on the work of Achak, A.; Bouhlal, A.; Daher, R.; Safouane, N.
2020 in Boletín de la Sociedad Matemática Mexicana

1.2.2 Paley-Wiener and Boas Theorems for the Quaternion Fourier Transform

DOI: https://doi.org/10.1007/s00006-013-0412-6

Original Abstract This paper establishes a real Paley-Wiener theorem to characterize the quaternion-valued functions whose quaternion Fourier transform has compact support by the partial derivative and also a Boas theorem to describe the quaternion Fourier transform of these functions that vanish on a neighborhood of the origin by an integral operator.

Acknowledgement
A machine generated summary based on the work of Fu, Yingxiong; Li, Luoqing
2013 in Advances in Applied Clifford Algebras

1.2.3 Paley-Wiener Theorems for the Two-Sided Quaternion Fourier Transform

DOI: https://doi.org/10.1007/s00006-016-0699-1

Original Abstract In this paper we establish new real Paley-Wiener theorems for the two-sided quaternion Fourier transform. Next, we prove the Roe's theorem in the context of the quaternion-valued functions. Finally we study the tempered distributions with spectral gaps.

Acknowledgement
A machine generated summary based on the work of Mejjaoli, Hatem
2016 in Advances in Applied Clifford Algebras

1.2.4 Bochner's Theorems in the Framework of Quaternion Analysis

DOI: https://doi.org/10.1007/978-3-0348-0603-9_5

Original Abstract Let σ(x) be a nondecreasing function, such that $\sigma(-\infty) = 0$, $\sigma(-\infty) = 1$ and let us denote by **B** the class of functions which can be represented by a Fourier–Stieltjes integral $f(t) = \int_{-\infty}^{\infty} e^{itx}\, d\sigma(x)$. The purpose of this chapter is to give a characterization of the class **B** and to give a generalization of the classical theorem of Bochner in the framework of quaternion analysis.

Acknowledgement
A machine generated summary based on the work of Georgiev, S.; Morais, J.
2013 in Trends in Mathematics

1.2.5 Bochner–Minlos Theorem and Quaternion Fourier Transform

DOI: https://doi.org/10.1007/978-3-0348-0603-9_6

Abstract-Summary
The first part of this chapter features certain properties of the asymptotic behaviour of the quaternion Fourier transform.

We introduce the quaternion Fourier transform of a probability measure, and we establish some of its basic properties.

Acknowledgement
A machine generated summary based on the work of Georgiev, S.; Morais, J.; Kou, K. I.; Spröߣig, W.
2013 in Trends in Mathematics

1.2.6 Hyers-Ulam Stability of Linear Quaternion-Valued Differential Equations with Constant Coefficients via Fourier Transform

DOI: https://doi.org/10.1007/s12346-022-00649-z

Abstract-Summary
We develop the Fourier transform approach to study the Hyers-Ulam stability of linear quaternion-valued differential equations with real coefficients and linear quaternion-valued even order differential equations with quaternion coefficients.

Extended:

Linear quaternion-valued (left type) differential equations with real coefficients and linear quaternion-valued even order (left type) differential equations with quaternion coefficients are investigated.

Introduction
Further, Chen and others [31] derived an explicit norm estimation for the quaternion-valued matrix exponential function in the sense of, which was used to show that the first-order linear equations are asymptotically stable and Hyers-Ulam's type stable and also showed that nth order equations are generalized Hyers-Ulam stable.

Chen and others [32] initially developed the tool of Laplace transform in [33] and studied Hyers-Ulam stability for linear differential equations in a quaternion-valued framework via Laplace transformation, which successfully transferred this well-known technique in linear ordinary differential equations (ODEs) to linear QDEs.

We transfer the analysis approach to study the Hyers-Ulam stability of Linear quaternion-valued differential equations with real coefficients and Linear quaternion-valued even order differential equations with quaternion coefficients.

Chen and others [31] firstly propose the asymptotical stability and the Hyers-Ulam stability concept of linear QDEs and apply the direct computation method via the newly established norm estimation of quaternion matrix exponential functions to prove the Hyers-Ulam stability results.

Hyers-Ulam Stability
We consider the Hyers-Ulam stability linear quaternion-valued (left type) differential equation with real coefficients and linear quaternion-valued even order differential equation with quaternion coefficients.

We obtain polynomials after the above quaternions are all real coefficient polynomials.

Conclusion
Linear quaternion-valued (left type) differential equations with real coefficients and linear quaternion-valued even order (left type) differential equations with quaternion coefficients are investigated.

To achieve our aim, the quaternion problem is converted into the corresponding complex representation problem.

Acknowledgement
A machine generated summary based on the work of Lv, Jiaojiao; Kou, Kit Ian; Wang, JinRong

2022 in Qualitative Theory of Dynamical Systems

1.2.7 Complex Interpolation of $L^p(C, H)$ Spaces with Respect to Cullen-Regular

DOI: https://doi.org/10.1007/s10473-022-0222-9

Abstract-Summary
The aim of this paper is to prove a new version of the Riesz-Thorin interpolation theorem on $L^p(C, H)$.

Two applications of the Riesz-Thorin theorem are presented.

Acknowledgement
A machine generated summary based on the work of He, Kaili; He, Jianxun; Lou, Zhenzhen

2022 in Acta Mathematica Scientia

2 Uncertainty

Machine Generated Keywords uncertainty principle, uncertainty, principle, two-side, two-side quaternion, quaternion fouri, QFT, fouri transform, fouri, principle quaternion, theorem, directional, two-side QFT, principle two-side, transform

2.1 Directional Uncertainty Principle for Quaternion Fourier Transform

DOI: https://doi.org/10.1007/s00006-009-0175-2

Abstract-Summary
This can be generalized to establish directional uncertainty principles in Clifford geometric algebras with quaternion subalgebras.

Acknowledgement
A machine generated summary based on the work of Hitzer, Eckhard M. S.

2009 in Advances in Applied Clifford Algebras

2.2 The Uncertainty Principle for the Two-Sided Quaternion Fourier Transform

DOI: https://doi.org/10.1007/s00009-017-1024-5

Original Abstract In this paper, we provide Heisenberg's inequality and Hardy's theorem for the two-sided quaternion Fourier transform.

Acknowledgement
A machine generated summary based on the work of El Haoui, Youssef; Fahlaoui, Said 2017 in Mediterranean Journal of Mathematics

2.3 A New Uncertainty Principle Related to the Generalized Quaternion Fourier Transform

DOI: https://doi.org/10.1007/s11868-021-00431-w

Abstract-Summary
Motivated by the treatment of the quaternion Fourier transform (QFT), the main purpose of the present paper is to characterize the spectrum of quaternion-value signals on the quaternion Fourier transform domain.

Introduction
Amrein and Berthier in [34] provided a different proof for that result.

The authors of [35] discussed the validity of the Benedicks-Amrein-Berthier type theorem in Clifford algebra using the Clifford–Fourier transform.

We intend to extend the validity of Benedicks-Amrein-Berthier's theorem in the case of the general two-sided QFT.

Quaternion Fourier Transforms
Ell [36] introduced the quaternion Fourier transform (QFT) as a hyper-complex transform which generalizes the classical Fourier transform (CFT) in the quaternion algebra setting.

Several important results and useful properties related to the extended transform appeared as modification and generalization of the corresponding ones of the CFT (more details can be found in [36, 23, 38, 15]).

They proposed a new form of the QFT for any two pure unit quaternions, called the general two-sided QFT [37, Definition 4.1], and then they discussed its geometric interpretation.

We recall the definition of the general two-sided QFT.

Benedicks–Amrein–Berthier Type Theorem
Due to the importance of UPs in signal and image analysis, many efforts have been placed to extend their validity in the context of the framework of the algebra of quaternions (see [23, 38, 15]).

Many theorems of the UPs such as Hardy, Beurling, Cowling-Price, Gelfand-Shilov and Miyachi have been generalized in the case of QFT (see for example, [38, 39]).

This section is devoted to the extension of Benedicks–Amrein–Berthier's uncertainty principle for the general two-sided QFT by virtue of the quaternion Fourier series expansion.

We begin by defining the notion of periodicity of a signal in the quaternion context.

Closing Remarks and Conclusion
The extension of this qualitative UP to the quaternionic algebra setting shows that a quaternionic 2D signal and its general two-sided QFT cannot both simultaneously have support of finite measure.

In the forthcoming work, we will investigate the validity of our main results for the space-time algebra of Minkowski.

Acknowledgement
A machine generated summary based on the work of El Haoui, Youssef; Zayed, Mohra

2021 in Journal of Pseudo-Differential Operators and Applications

2.4 *Miyachi's Theorem for the Quaternion Fourier Transform*

DOI: https://doi.org/10.1007/s00034-019-01243-6

Abstract-Summary
The quaternion Fourier transform (QFT) satisfies some uncertainty principles similar to the Euclidean Fourier transform.

Extended:

The quaternion Fourier transform (QFT) is a non-trivial extension of the real and complex classical Fourier transform to the algebra of the quaternions.

Introduction
Bülow [39] investigated the important properties of the two-sided QFT for real signals and applied it to signal and image processing.

In [10], the authors generalized a component-wise UP for the right-sided QFT.

The directional UP related to the two-sided QFT was proposed in [16].

No such uncertainty principle for the QFT (one-sided or two-sided) had been established.

Other UPs: Heisenberg, Hardy [38], and Beurling [40], have been extended for the two-sided QFT.

We will obtain, in a different way, by the main result of Miyachi, the same UPs of Cowling–Price and Hardy in the QFT domain.

Miyachi's Theorem
We prove Miyachi's theorem for the quaternion Fourier transform, and Cowling–Price's uncertainty principle for the QFT.

Conclusions

Based on some obtained results of the two-sided QFT and one technical lemma of complex analysis, a generalization of Miyachi's uncertainty principle associated with the QFT was proposed.

The extension of these qualitative UPs to the quaternionic algebra framework shows that a quaternionic 2D signal f and its QFT cannot both simultaneously decrease very rapidly.

The QFT has proved to be a very significant tool for applications in color image processing [41], quantum mechanics, engineering, signal processing, optics, etc. Apart from their importance to pure mathematics, our results are also relevant to applied mathematics and signal processing.

Acknowledgement

A machine generated summary based on the work of El Haoui, Youssef; Fahlaoui, Said 2019 in Circuits, Systems, and Signal Processing

2.5 *Tighter Uncertainty Principles Based on Quaternion Fourier Transform*

DOI: https://doi.org/10.1007/s00006-015-0579-0

Original Abstract The quaternion Fourier transform (QFT) and its properties are reviewed in this paper. Under the polar coordinate form for quaternion-valued signals, we strengthen the stronger uncertainty principles in terms of covariance for quaternion-valued signals based on the right-sided quaternion Fourier transform in both the directional and the spatial cases. We also obtain the conditions that give rise to the equal relations of two uncertainty principles. Examples are given to verify the results.

Acknowledgement

A machine generated summary based on the work of Yang, Yan; Dang, Pei; Qian, Tao 2015 in Advances in Applied Clifford Algebras

2.6 *A Modified Uncertainty Principle for Two-Sided Quaternion Fourier Transform*

DOI: https://doi.org/10.1007/s00006-015-0617-y

Abstract-Summary

This paper proposes a new uncertainty principle for the two-sided quaternion Fourier transform.

This uncertainty principle describes that the spread of a quaternion-valued function and its two-sided quaternion Fourier transform (QFT) are inversely proportional.

Acknowledgement
A machine generated summary based on the work of Bahri, Mawardi
2015 in Advances in Applied Clifford Algebras

2.7 *Donoho–Stark's Uncertainty Principle for the Quaternion Fourier Transform*

DOI: https://doi.org/10.1007/s40590-019-00251-5

Abstract-Summary
The (right-sided) quaternion Fourier transform satisfies some uncertainty principles in a similar way to the Euclidean Fourier transform.

Introduction
The uncertainty principle says that a function and its transform both cannot concentrate on small sets.

This paper focuses on studying different uncertainty principles for the quaternion Fourier transform by following the procedures for similar transforms such as the Fourier transform (the classical setting); we refer the reader to the excellent survey article by Folland and Sitaram [42] and also the monograph by Thangavelu [43].

We study some uncertainty principles for the right quaternion Fourier transform.

We generalize the Donoho–Stark principle to the less known setting of the quaternion Fourier transform.

Recall that Donoho and Stark [44] paid attention to the support of functions and gave quantitative uncertainty principles for the Fourier transforms.

To describe our results, we first need to introduce some facts about harmonic analysis related to the (right-sided) quaternion Fourier transform (QFT).

Donoho–Stark Uncertainty Principles for the Quaternion Fourier Transform
It has become popular to generalize the integral transforms from real and complex numbers to quaternion algebra to study higher dimensions such as the quaternion Fourier transform (QFT).

It is not a surprise that uncertainty principles also hold for the QFT.

The notion of 'close to zero' is formulated.

Acknowledgement
A machine generated summary based on the work of Abouelaz, A.; Achak, A.; Daher, R.; Safouane, N.
2019 in Boletín de la Sociedad Matemática Mexicana

2.8 *Beurling's Theorem for the Quaternion Fourier Transform*

DOI: https://doi.org/10.1007/s11868-019-00281-7

Abstract-Summary
The two-sided quaternion Fourier transform satisfies some uncertainty principles similar to the Euclidean Fourier transform.

Extended:

We will consider Benedicks–Amrein–Berthier's UP for the quaternionic Fourier transform.

Introduction
In harmonic analysis, the uncertainty principle states that a non zero function and its Fourier transform cannot both be sharply localized.

We further emphasize that our generalization is non-trivial because the multiplication of quaternions and the quaternion Fourier kernel are both non-commutative.

The Two-Sided Quaternion Fourier Transform
There are three different types of QFT, the left-sided QFT, the right-sided QFT, and two-sided QFT [45].

We now review the definition and some properties of the two-sided QFT.

Conclusions
We derive Beurling's UP for the QFT, which is a qualitative version of UP.

It shows that it is not possible for a nonzero quaternionic 2D signal and its QFT to both decrease very rapidly.

With regard to the decreasing property, we propose three variants for the Beurling's UP in the QFT domain: Hardy's UP, Cowling–Price's UP and Gelfand–Shilov's UP.

Acknowledgement
A machine generated summary based on the work of El Haoui, Youssef; Fahlaoui, Said 2019 in Journal of Pseudo-Differential Operators and Applications

2.9 *Uncertainty Principle for the Two Sided Quaternion Windowed Fourier Transform*

DOI: https://doi.org/10.1007/s11868-019-00283-5

Introduction
Uncertainty principles are mathematical results that give limitations on the simultaneaous concentration of a function and its quaternion Fourier transform [46, 42].

They have implications in two main areas: quantum physics and signal analysis.

In [47], Hahn constructed a quaternion Wigner–Ville distribution (QWVT) and radar quaternion ambiguity (QAF) function of 2D quaternionic signals which is in fact closely related to the QWFT.

Our purpose in this work is to prove three Heisenberg uncertainty principles for the QWFT and QWVT and QAF.

Basic Notions and Preliminaries Results
The quaternion algebra was formally introduced by Hamilton in 1843, and it is a generalization of complex numbers.

In the following we briefly introduce the two sided QFT.

Uncertainty Principle
We prove Lieb's uncertainty principle for the two-sided quaternion windowed Fourier transform.

The proof is identical to that given by [48].

Acknowledgement
A machine generated summary based on the work of Brahim, Kamel; Tefjeni, Emna 2019 in Journal of Pseudo-Differential Operators and Applications

3 Signal Processing

Machine Generated Keywords sample formula, sample, rate, quaternionic, parameter, transmission, conventional, formula, canonical transform, linear canonical, canonical, signal, application quaternion, key, change

3.1 *Sampling Formulas for Non-bandlimited Quaternionic Signals*

DOI: https://doi.org/10.1007/s11760-021-02110-1

Abstract-Summary
Sampling theorems for certain types of non-bandlimited quaternionic signals are proposed.

Extended:

By the generalized sinc functions, we will first establish the spaces of NBL quaternionic signals, and then the sampling theorems for two classes of NBL quaternionic signals are established.

It will be a possible direction for our future work.

Introduction
Enlightened by [49, 50], the authors in [51] structured a two-parameter ladder-shaped filter in the linear canonical transform domain and established sampling theorems for NBL real signals in the linear canonical transform sense.

Using quaternion prolate spheroidal wave functions, sampling theorems for 2D BL quaternionic signals were first derived in paper [52].

Sampling theorems for 2D BL quaternionic signals associated with right-sided QFT and QLCT were proposed in paper [53].

In paper [54], the authors have derived sampling theorems of 2D BL quaternionic signals under various QFTs and QLCTs.

By the generalized sinc functions, we will first establish the spaces of NBL quaternionic signals, and then the sampling theorems for two classes of NBL quaternionic signals are established.

We propose four types of sampling formulas for the NBL quaternionic signals associated with the QLCT and QFT, respectively.

Preliminaries

This section is devoted to the exposition of basic preliminary material which we use extensively throughout of this paper.

Based on the concept of quaternion, many types of QFTs and QLCTs [55] have been introduced.

Sampling Theorems for NBL Quaternionic Signals

We deal with the quaternionic signals which are non-bandlimited in the frequency domain.

We introduce a function space consisting of NBL quaternionic signals with special spectrum properties associated with a ladder-shaped filter of two parameters.

Experimental Results

We also compare our methods with the sampling formula for quaternionic signals in paper [54].

It is seen that the proposed sampling formulas can be applied effectively in the reconstruction of non-bandlimited signals.

It can be seen that the sampling formulas for non-bandlimited signals outperform the sampling formula for bandlimited signals [54].

For quaternionic signal f(x), 250 sampling points are needed for reconstruction, while 150 sampling points are enough for the reconstruction of the quaternionic signal g(x).

Conclusion and Discussion

These sampling formulas are applicable for certain types of the non-bandlimited quaternionic signals.

According to the quaternion Riemann-Lebesgue lemma [55], a majority of non-bandlimited quaternionic signals satisfy the requirement of the proposed theorems approximately.

Quaternions are gaining popularity in signal and image processing (see e.g., [56, 57]).

There is a need to develop the 2D sampling series of non-bandlimited quaternionic signals to provide an effective tool in image processing.

Acknowledgement

A machine generated summary based on the work of Hu, Xiaoxiao; Kou, Kit Ian 2022 in Signal, Image and Video Processing

3.2 *Sampling Formulas for 2D Quaternionic Signals Associated with Various Quaternion Fourier and Linear Canonical Transforms*

DOI: https://doi.org/10.1631/fitee.2000499

Abstract-Summary
The main purpose of this paper is to study different types of sampling formulas of quaternionic functions, which are bandlimited under various quaternion Fourier and linear canonical transforms.

If the quaternionic function is bandlimited to a rectangle that is symmetric about the origin, then the sampling formulas under various quaternion Fourier transforms are identical.

If this rectangle is not symmetric about the origin, then the sampling formulas under various quaternion Fourier transforms are different from each other.

Using the relationship between the two-sided quaternion Fourier transform and the linear canonical transform, we derive sampling formulas under various quaternion linear canonical transforms.

Acknowledgement
A machine generated summary based on the work of Hu, Xiaoxiao; Cheng, Dong; Kou, Kit Ian

2022 in Frontiers of Information Technology & Electronic Engineering

3.3 *Quaternionic Spectral Analysis of Non-stationary Improper Complex Signals*

DOI: https://doi.org/10.1007/978-3-0348-0603-9_3

Abstract-Summary
The signals considered are one-dimensional (1D), complex-valued, with possible correlation between the real and imaginary parts, i.e., improper complex signals.

This does not provide a geometric analysis of the complex improper signal.

We propose another approach for the analysis of improper complex signals based on the use of a 1D Quaternion Fourier Transform (QFT).

In the case where complex signals are non-stationary, we investigate the extension of the well-known 'analytic signal' and introduce the quaternion-valued 'hyperanalytic signal'.

As with the hypercomplex two-dimensional (2D) extension of the analytic signal proposed by Bülow in 2001, our extension of analytic signals for complex-valued signals can be obtained by the inverse QFT of the quaternion-valued spectrum after suppressing negative frequencies.

Analysis of the hyperanalytic signal reveals the time-varying frequency content of the corresponding complex signal.

The hyperanalytic signal can be seen as the exact counterpart of the classical analytic signal, and should be thought of as the very first and simplest quaternionic time-frequency representation for improper non-stationary complex-valued signals.

Acknowledgement
A machine generated summary based on the work of Bihan, Nicolas Le; Sangwine, Stephen J.
2013 in Trends in Mathematics

3.4 *Many-Parameter Quaternion Fourier Transforms for Intelligent OFDM Telecommunication System*

DOI: https://doi.org/10.1007/978-3-030-39162-1_8

Abstract-Summary
We aim to investigate the superiority and practicability of many-parameter quaternion Fourier transforms (MPQFT) from the physical layer security (PHY-LS) perspective.

We propose novel Intelligent OFDM-telecommunication system (Intelligent-OFDM-TCS), based on MPFT.

Each MPQFT depends on a finite set of independent parameters (angles), which could be changed independently one from another.

When parameters are changed, the multi-parametric transform is also changed taking the form of a set of known (and unknown) orthogonal (or unitary) transforms.

Extended:

We have shown a new unified approach to the many-parametric representation of complex and quaternion Fourier transforms.

Simulation results show that the proposed Intelligent OFDM-TCS have better performance than the conventional OFDM system based on the DFT against eavesdropping.

Introduction
Most of the data transmission systems nowadays use orthogonal frequency division multiplexing telecommunication system (OFDM-TCS) based on the discrete Fourier transform (DFT).

OFDM divides the available spectrum into a number of parallel orthogonal sub-carriers and each sub-carrier is then modulated by a low rate data stream at a different carrier frequency.

There are a number of candidates for orthogonal function sets used in the OFDM-TCS: discrete wavelet sub-carriers [58], Golay complementary sequences [59], Walsh functions [60], pseudo random sequences [61].

If, nevertheless, this key were found by the enemy in a cyber attack, then the system could change values of the working parameters for rejecting the enemy attack.

Knowing this key is necessary to enter into the Intelligent OFDM TCS.

Such a system can also defend itself by changing the values of the working parameters and crypto key according to some law that is known for the transmitter and receiver in advance.

Quaternion Fourier Transforms

Before defining the quaternion Fourier transform, we briefly outline its relationship with Clifford Fourier transformations.

Quaternions and Clifford hypercomplex numbers were first simultaneously and independently applied to quaternion-valued Fourier and Clifford-valued Fourier transforms by Labunets [62] and Sommen [63, 64], respectively, in 1981.

Labunets, Rundblad-Ostheimer and Astola [65, 66–67] used the classical and number theoretical quaternion Fourier and Fourier-Clifford transforms for fast invariant recognition of 2D, 3D and nD color and hyperspectral images, defined on Euclidean and non-Euclidean spaces.

These publications give useful interpretation of quaternion and Cliffordean Fourier coefficients: they are relative quaternion- or Clifford-valued invariants of hyperspectral images with respect to Euclidean and non-Euclidean rotations and motions of physical and hyperspectral spaces.

Due to the non-commutative property of quaternion multiplication, there are two different types of quaternion Fourier transforms (QFTs).

Fractional and Many-Parameter Quaternion Fourier Transforms

We can define fractional and many parameter quaternion Fourier transforms.

Due to the non-commutative property of quaternion multiplication, there are left- and right-sided transforms (LS-FrQFTs, LS-MPQFTs and RS-FrQFTs, RS-MPQFTs).

Anti-eavesdropping: Bob and Alice vs. Eve

This model is composed of two legitimate users, named Alice and Bob, while the passive eavesdropper named Eve attempts to eavesdrop the information.

A legitimate user (Alice) transmits her confidential messages to a legitimate receiver (Bob), while Eve is trying to eavesdrop Alice's information.

The legitimate transmitter/receiver (Alice/Bob) and eavesdropper (Eve) use identical parameters of Intel-OFDM-TCS which remain constant over several time slots.

Bob and Eve will have the same instruments to decode the received message.

Conclusions

We have shown a new unified approach to the many-parametric representation of complex and quaternion Fourier transforms.

Defined representation of many-parameter quaternion Fourier transforms (MPQFTs) depend on a finite set of free parameters, which could be changed independently of one another.

We develop novel Intelligent OFDM-telecommunication systems based on fractional and multi-parameter Fourier transforms and shown their superiority and practicability from the physical layer security viewpoint.

Acknowledgement
A machine generated summary based on the work of Labunets, Valeriy G.; Ostheimer, Ekaterina
2020 in Advances in Intelligent Systems and Computing

3.5 *Intelligent OFDM Telecommunication Systems Based on Many-Parameter Complex or Quaternion Fourier Transforms*

DOI: https://doi.org/10.1007/978-3-030-39216-1_13

Abstract-Summary
We propose novel Intelligent quaternion OFDM-telecommunication systems based on a many-parameter complex and quaternion transform (MPFT).

Each MPFT depends on a finite set of independent parameters (angles).

The concrete values of parameters are the specific "key" for entry into OFDM-TCS.

Scanning of this space for finding the "key" (the concrete values of parameters) is a hard problem.

Extended:

We propose a simple and effective anti-eavesdropping and anti-jamming Intelligent OFDM system, based on many-parameter complex or quaternion Fourier transforms (MPFTs).

We aim to investigate the superiority and practicability of MPFTs from the physical layer security (PHY-LS) perspective.

Introduction
Ensuring information security is of paramount importance for wireless communications.

The layered design architecture with a transparent physical layer leads to a loss in security functionality [68], especially for wireless communication scenarios where a common physical medium is always shared by legitimate and non-legitimate users.

Exploiting physical layer characteristics for secure transmission has become an emerging hot topic in wireless communications [69, 70, 71–72].

As the physical-layer transmission adversaries can blindly estimate parameters of OFDM signals, traditional upper-layer security mechanisms cannot completely address security threats in wireless OFDM systems.

Physical layer security, which targets communications security at the physical layer, is emerging as an effective complement to traditional security strategies in securing wireless OFDM transmission.

The physical layer security of OFDM systems over wireless channels was investigated from an information-theoretic perspective in [73].

Intelligent Complex and Quaternion OFDM TCS

There are a number of candidates for orthogonal function sets used in the OFDM-TCS: discrete wavelet sub-carriers [74, 58], Golay complementary sequences [75, 76], Walsh functions [60, 77], and pseudo random sequences [78, 79].

Intelligent-OFDM TCS can be described as a dynamically reconfigurable TCS that can adaptively regulate its internal parameters as a response to changes in the surrounding environment.

One of the most important capacities of Intelligent OFDM systems is their capability to optimally adapt their operating parameters based on observations and previous experiences.

We propose a simple and effective anti-eavesdropping and anti-jamming Intelligent OFDM system, based on a many-parameter transform.

The main advantage of using MPFT in OFDM TCS is that it is a very flexible anti-eavesdropping and anti-jamming Intelligent OFDM TCS.

Knowing this digital key is necessary to enter into the Intelligent OFDM TCS.

Conclusions

We proposed a novel Intelligent OFDM-telecommunication systems based on a new unified approach to the many-parametric representation of complex and quaternion Fourier transforms.

When parameters are changed, the multi-parametric transform is changed too taking the form of a set known (and unknown) orthogonal (or unitary) transforms.

The main advantage of using MPFT in OFDM TCS is that it is a very flexible system allowing to construct Intelligent OFDM TCS for electronic warfare (EW).

Suppressor aims to "reduce the effectiveness" of enemy forces, including command and control and their use of weapons systems, and targets enemy communications and reconnaissance by changing the "quality and speed" of information processes.

In order to protect corporate privacy and sensitive client information against the threat of electronic eavesdropping and jamming, the protector uses intelligent OFDM-TCS, based on MPFTs.

A legitimate user (Alice) transmits her confidential messages to a legitimate receiver (Bob), while Eve will be trying to eavesdrop Alice's information.

Acknowledgement

A machine generated summary based on the work of Labunets, Valeriy G.; Ostheimer, Ekaterina

2020 in Advances in Intelligent Systems and Computing

3.6 Quaternion Fourier Descriptors for the Preprocessing and Recognition of Spoken Words Using Images of Spatiotemporal Representations

DOI: https://doi.org/10.1007/s10851-007-0004-y

Abstract-Summary
This paper presents an application of the quaternion Fourier transform for the preprocessing for neural-computing.

This kind of images are then convolved in the quaternion Fourier domain with a quaternion Gabor filter for the extraction of features.

The improvements in the classification rate of the neural network classifiers are very encouraging which amply justify the preprocessing in the quaternion frequency domain.

This work also suggests the application of the quaternion Fourier transform for other image processing tasks.

Acknowledgement
A machine generated summary based on the work of Bayro-Corrochano, Eduardo; Trujillo, Noel; Naranjo, Michel

2007 in Journal of Mathematical Imaging and Vision

4 Image Processing

Machine Generated Keywords color, filter, texture, color image, invariant, power, QFT, noise, separate, model, parameter, hardy, index, edge detection, call quaternion

4.1 The Quaternion-Fourier Transform and Applications

DOI: https://doi.org/10.1007/978-3-030-06161-6_15

Abstract-Summary
To filter color images, a new approach is implemented recently which uses hypercomplex numbers (called quaternions) to represent color images and uses the quaternion fourier transform for filtering.

The quaternion Fourier transform has been widely employed in colour image processing.

Introduction
We will consider the real Paley-Wiener theorem for the quaternion Fourier transform (QFT) which is a nontrivial generalization of the real and complex Fourier transform (FT) to quaternion algebra.

The four components of the QFT separate four cases of symmetry in real signals instead of only two in the complex FT.

The QFT plays an important role in the representation of signals and transforms a quaternion 2D signal into a quaternion-valued frequency domain signal.

Preliminaries
We recall the following important property of the QFT.

For more properties and details, we refer to [80, 15].

Real Paley-Wiener Theorems for the Quaternion-Fourier Transform
We consider the functions vanishing outside a ball, which is the Paley-Wiener-Type Theorem.

We consider the functions vanishing on a ball, which is the Boas-Type Theorem.

Acknowledgement
A machine generated summary based on the work of Li, Shanshan; Leng, Jinsong; Fei, Minggang

2019 in Lecture Notes of the Institute for Computer Sciences, Social Informatics and Telecommunications Engineering

4.2 Quaternion Fourier Transform: Re-tooling Image and Signal Processing Analysis

DOI: https://doi.org/10.1007/978-3-0348-0603-9_1

Original Abstract Quaternion Fourier transforms (QFTs) provide expressive power and elegance in the analysis of higher-dimensional linear invariant systems. But, this power comes at a cost – an overwhelming number of choices in the QFT definition, each with consequences. This chapter explores the evolution of QFT definitions as a framework from which to solve specific problems in vector-image and vector-signal processing.

Acknowledgement
A machine generated summary based on the work of Ell, Todd Anthony

2013 in Trends in Mathematics

4.3 A Robust Color Edge Detection Algorithm Based on the Quaternion Hardy Filter

DOI: https://doi.org/10.1007/s10473-022-0325-3

Abstract-Summary
This paper presents a robust filter called the quaternion Hardy filter (QHF) for color image edge detection.

The QHF can be capable of color edge feature enhancement and noise resistance.

The quaternion analytic signal, which is an effective tool in color image processing, can also be produced by quaternion Hardy filtering with specific parameters.

Based on the QHF and the improved Di Zenzo gradient operator, a novel color edge detection algorithm is proposed; importantly, it can be efficiently implemented by using the fast discrete quaternion Fourier transform technique.

Acknowledgement
A machine generated summary based on the work of Bi, Wenshan; Cheng, Dong; Liu, Wankai; Kou, Kit Ian

2022 in Acta Mathematica Scientia

4.4 Denoising Color Images Based on Local Orientation Estimation and CNN Classifier

DOI: https://doi.org/10.1007/s10851-019-00942-8

Abstract-Summary
A structure-adaptive vector filter for removal of impulse noise from color images is presented.

A color image is represented in quaternion form, and then, the quaternion Fourier transform is used to compute the orientation of the pattern in a local neighborhood.

To further improve the denoising effect, a deep convolutional neural network is employed to detect impulse noise in color images and integrated into the proposed denoising framework.

The experimental results exhibit the effectiveness of the proposed denoiser by showing significant performance improvements both in noise suppression and in detail preservation, compared to other color image denoising methods.

Extended:

A color image can be represented as different quaternion forms.

Introduction
Numerous filtering methods for impulse noise reduction have been developed.

For color images, impulse noise removal techniques mainly include component-wise methods, vector-processing methods, and other techniques [81].

Vector filtering techniques, which treat color images as vector fields, are often more effective than component-wise methods, due to the strong correlation between color channels.

These algorithms [82, 83, 84–85] train a NN to be an impulse noise detector, which is then used to detect noisy pixels in images, and finally employ different filtering mechanisms to restore the noisy pixels detected.

Most machine learning and NN-based methods are designed for grayscale image denoising and few are developed for color image denoising.

The authors Jin and others [86] proposed a highly effective structure-adaptive VMF for removing impulse noise from color images.

To further boost denoising performance, an effective noise detection mechanism is integrated into the structure-adaptive vector filter.

Proposed Method
The noise classifier CNN [87] takes a color image as the input and outputs the label vectors of all pixels.

The label vector of each pixel not only denotes whether the color pixel is corrupted by impulse noise but also further identifies whether individual channels are noisy or not.

Three matrixes are generated by three convolutional filters of size $3 \times 3 \times 64$, each of which gives the probabilities that all the input pixels in a color channel are corrupted by impulse noise.

In label images, each pixel is a 3D column vector, where each element is 1 or 0 denoting whether the corresponding color channel is corrupted by impulse noise or not.

The trained classifier CNN is used to analyze the noise property of noisy color images and outputs label vector images.

Experiments
To evaluate the performance of a denoising method, simulated impulse noise is needed to add to noise-free color images.

We first evaluate the performance of the CNN noise classifier, then test the effect of the proposed orientation detection mechanism on denoising performance, and finally compare the proposed denoiser with other color image denoising methods.

We evaluate the performance of the CNN classifier by comparing it with other machine learning and neural network methods: SVM-based detector with center weighted median filter (SVM-MF) [88] and ANN-based impulse detector with edge-preserving regularization (ANN-EPR) [82].

From these tables, one can observe that the proposed method consistently provides robust performance and significantly outperforms other methods used for comparison in terms of all the four objective criteria on all test images in various noise contamination levels.

One can observe that the proposed denoiser removes noise sufficiently and at the same time preserves image textures and details well, yielding the best visual effect compared to other methods used in the comparison.

Conclusion
A structure-preserving vector filter for suppressing color impulse noise is proposed.

The proposed method is based on a local orientation estimation and a CNN noise classifier.

The dominant orientation of a local color pattern is estimated by computing the orientation of the power spectrum distribution of the quaternion Fourier transform of the local pattern.

The experimental results show the superiority of the proposed method in both noise removal and structure preservation, compared to other state-of-the-art color image denoising methods.

Acknowledgement
A machine generated summary based on the work of Jin, Lianghai; Song, Enmin; Zhang, Wenhua

2020 in Journal of Mathematical Imaging and Vision

4.5 Rotation Invariant Features for Color Texture Classification and Retrieval Under Varying Illumination

DOI: https://doi.org/10.1007/s10044-011-0207-0

Abstract-Summary
This article proposes a new quaternion-based method for rotation invariant color texture classification under illumination variance with respect to direction and spectral band.

These signatures are proved to be invariant under image rotation and illumination rotation.

The robustness of different color spaces against varying illumination in color Texture classification with 45 samples of 15 outex texture classes are examined.

Comparative results show that the proposed method is efficient in rotation invariant texture classification.

Extended:

This article proposes a Quaternion-based rotation invariant classification method for color textures with illumination changes.

Introduction
Early methods for texture classification focus on the statistical, structural, model based and signal processing analysis of texture images.

Kashyap and Khotanzad [89] were among the first researchers to study rotation-invariant texture classification using a circular autoregressive model.

Many Gabor and wavelet-based algorithms were also proposed for rotation invariant texture classification.

The direction of illumination has a binding influence on the directional characteristics of texture images.

The image forming process under directed illumination of texture acts as directional filter.

Among the work reported in texture classification only very few include the effect of illumination in grayscale images [90].

The color of the image can aid the process of texture classification and retrieval [91].

This article proposes a Quaternion-based rotation invariant classification method for color textures with illumination changes.

Rotation Invariant Texture Classification
Further quaternion texture analysis, quaternion singular value decomposition and quaternion principal component analysis are implemented and applied to several applications, such as segmentation of color images [92].

The quaternion representation of color is shown to be effective in the context of color texture analysis in digital color images [93].

The advantage of the quaternion representation for color image I(m, n), as proposed by Sangwine, combines a color 3-tuple (RGB or LMS) of I(m, n) into a single hyper complex number.

Any rotation invariant texture signature in the quaternion domain will be invariant to both orientation and illumination changes.

The idea of computing the quaternion Fourier transform (QFT) of a color image has only recently been realized.

They utilized a hyper complex Fourier transform because of its symmetry properties when applied to real-valued images but they haven't considered application of the transform to invariant texture analysis of color images.

Experiment and Results
This approach for the texture classification uses the peak distribution norm vector (PDNV) of the quaternion derived from the radial plot as rotation invariant texture signature.

For some selected 15 textures, the proposed method yields a classification accuracy of 97.46 and 92.07, 83.55, and 76.46% are achieved by Gray FFT, Log Polar, and Gabor methods, respectively, for the outex data set.

Precision versus Recall have been plotted for the conventional GLCM [94, 95], moment invariants [96], gray FFT methods and the proposed QFFT-based method.

Depending on which color space is chosen, very different classification performances are observed using the same texture features and the same experimental image data so the choice of color space for texture analysis is crucial.

The QDFT or its FFT implementation QFFT requires fewer real multiplications than three-complex DFT/FFTs and hence is more efficiently computed for a color image, as well as requiring less memory.

Conclusion
A Novel Quaternion-based rotation-invariant texture classification and retrieval has been proposed.

The rotation-invariant texture signatures, and the peak distribution norm vector have been constructed from the radial plot of the Quaternion Fourier transform and were further used for classification.

The emphasis of this paper has been on extracting the rotation-invariant texture features.

Acknowledgement
A machine generated summary based on the work of Sathyabama, B.; Anitha, M.; Raju, S.; Abhaikumar, V.
2011 in Pattern Analysis and Applications

4.6 *Quaternion-Based Texture Analysis of Multiband Satellite Images: Application to the Estimation of Aboveground Biomass in the East Region of Cameroon*

DOI: https://doi.org/10.1007/s10441-018-9317-z

Abstract-Summary
Several approaches using quaternion numbers to handle and model multiband images in a holistic manner were introduced.

The quaternion Fourier transform can be efficiently used to model texture in multidimensional data such as color images.

We propose a texture-color descriptor based on the quaternion Fourier transform to extract relevant information from multiband satellite images.

We propose a new multiband image texture model extraction, called FOTO++, in order to address biomass estimation issues.

Color texture descriptors are extracted thanks to a discrete form of the quaternion Fourier transform, and finally the support vector regression method is used to deduce biomass estimation from texture indices.

Our texture features are modeled using a vector composed with the radial spectrum coming from the amplitude of the quaternion Fourier transform.

The results show that our model is more robust to acquisition parameters than the classical Fourier Texture Ordination model (FOTO).

These results highlight the potential of the quaternion-based approach to study multispectral satellite images.

Extended:

The quaternion Fourier transform treats the signals as vector fields, and generalizes the conventional Fourier transform.

The quaternion Fourier spectrum, as in the complex Fourier transform, can be divided into four quadrants, and in each quadrant the coordinates correspond to exactly one of the four combinations of positive and negative horizontal and vertical spatial frequencies.

We are dealing with multispectral satellite data acquired using passive sensors.

Introduction

Traditional methods (such as destructive sampling and allometric equations) based on forest survey data constitute the most accurate way to estimate the biomass of a specific area, but they remain time-consuming and labor intensive (Ketterings and others [97]; Chave and others [98, 99]; Basuki and others [100]; Saatchi and others [101]; Feldpausch and others [102]; Vieilledent and others [103]; Fayolle and others [104]; Hunter and others [105]; Kearsley and others [106]; Mitchard and others [107; Ekoungoulou and others [108], [109]; Picard and others [110]; Gaia and others [111]; Re and others [112]).

Two main approaches are available to extract relevant information from remote sensing data for aboveground biomass estimation: the reflectance-based approach and the image texture-based approach.

One major limitation encountered in the reflectance-based analysis of satellite image data is due to the fact that a reflectance is always saturated when a forest has a high value of biomass.

These methods rely on the assumption that estimating biomass from satellite data can be tackled by image texture classification.

One of the most popular methods to address aboveground forest biomass estimation is the FOurier-based Texture Ordination (FOTO) (Couteron [113]; Couteron and others [114]).

Despite of numerous works dealing with multiband image processing, only few methods based on multispectral satellite image analysis have been proposed to address biomass issues.

Texture-Based Analysis Image for Forest Biomass Estimation: The FOTO Method

FOurier-based Texture Ordination (FOTO) is an aboveground biomass estimation method performed on data obtained from monochromatic satellite canopy image texture analysis.

The basic principle is to link canopy grain to image texture features derived from Fourier spectrum analysis.

After feature extraction, for each sub-image, 2D Fourier transform periodograms are calculated to characterize textural image properties.

It takes into account the spatial frequency information contained in each sub-image and provides the so-called R-spectrum.

The FOTO method can handle only monochromatic (panchromatic data) images for texture analysis, although multispectral information (e.g. visible colors combined with infrared) can be of great importance.

When the spectral bands are pooled together, they provide one multispectral image that needs to be addressed using multiband image processing methods.

The choice of an appropriate color space can improve significantly the quality of the image processing methods.

Color Image Processing and Quaternion Algebra

In these color spaces, the hue and the saturation components represent the color information, while the intensity component describes either the amount of light (the brightness and the value component) or the amount of white (the lightness component) of the image.

True color images are obtained by putting multispectral satellite images in the RGB color space with the red, the green and the blue bands assigned to red, green and blue channels respectively.

In the case where the near infrared (NIR) band is used jointly with the red and the green bands and are assigned to the red, the green and the blue channels respectively, we have a pseudo-natural color image.

The spatio-frequential approaches aim to characterize the color and texture of multiband images in the frequency domain because it is possible to express all the information present in the image using just a small number of coefficients.

Texture-Based Analysis of Color Images for Biomass Estimation: The FOTO++ Method

Powerful tools for multiband image processing have followed: Quaternion Fourier Transform (Sangwine [93]; Sangwine and Ell [115]; Pei and others [45]; Ell and Sangwine [116]), Auto-correlation (Sangwine and Ell [115]; Moxey and others [117]), Cross-correlation (Sangwine and Ell [115]), Convolution (Pei and others [45]), Color Texture analysis (Shi and Funt [118]).

The aim of FOTO++ is to model texture in multiband images thanks to the quaternion Fourier transform.

A classical way to extend the two-dimensional Fourier transform to quaternion images consists in replacing the complex number i with a pure quaternion square root of -1, denoted by μ.

We choose the left single-axis form of the quaternion Fourier transform and its discrete form as proposed by Sangwine [93] for color image processing.

We apply the zero-padding process to function f before taking the quaternion Fourier transform to deal with image border issues.

We characterize the color image texture pattern by performing a periodogram analysis of the DQFT spectrum as delineated by Mugglestone and Renshaw [119] for the classical Fourier transform.

Experimental Results

In order to precisely evaluate the contribution of the FOTO++ model in above-ground biomass estimation, we calculated and compared the performances of the classical FOTO model, FOTO with SVR used as prediction tool model (Tapamo and others [120]) and the FOTO++ model.

We have computed the texture indices of each image of the simulated dataset by means of the FOTO++ model.

We calculated and compared RMSE values and Willmott's model performance index of FOTO-SVR and FOTO++ without taking into account the filtering step.

In this last subsection, we evaluated how a filtering step using the Nagao-median filter contributes to biomass estimate accuracy in the FOTO++ model.

We can conclude, based on these experiments, that color-texture indices extracted with the quaternion Fourier transform combined with machine learning tools and with the use of filtering techniques can significantly improve the accuracy of biomass estimates compared with the classical FOTO model.

Conclusion
The problem of estimating biomass from satellite data can be viewed as image texture analysis and classification.

One such approach is the FOurier-based Texture Ordination (FOTO) method proposed by Couteron 113 for aboveground forest biomass estimation.

By introducing quaternions, the quaternion Fourier transform allows multidimensional data such as multiband images to be processed efficiently and easily.

After this preprocessing stage, multiband texture indices were extracted using the discrete form of the quaternion Fourier transform, and finally the support vector regression method was used to derive biomass estimations from these multiband texture indices.

We used quaternion-based tools to extract useful texture features from multiband satellite images.

When applying SVR on quaternion-based color-texture, an improvement of 24.02% is observed with on average 82.70% of biomass value free of error.

Quaternion-based color-textured and filtering are valuable tools that can be used to improve biomass value estimates.

Acknowledgement
A machine generated summary based on the work of Djiongo Kenfack, Cedrigue Boris; Monga, Olivier; Mpong, Serge Moto; Ndoundam, René

2018 in Acta Biotheoretica

5 Watermarking

Machine Generated Keywords watermark, watermarking, color image, embed, color, digital, copyright, scheme, watermarke, encryption, channel, algorithm, protection, fractional, utilize

5.1 DWT-DQFT-Based Color Image Blind Watermark with QR Decomposition

DOI: https://doi.org/10.1007/978-3-030-89137-4_15

Abstract-Summary
In order to improve the performance of color image digital watermarking, a watermarking algorithm based on integration of Discrete Wavelet Transform (DWT) and Discrete Quaternion Fourier Transform (DQFT) combined with QR matrix decomposition is proposed.

Introduction

A lot of algorithms of digital watermarking have been proposed, mainly including embedding the watermark directly in the pixel values of host images [121], or altering the coefficients in the transform domain, such as the Fourier transform, discrete cosine transform [122, 123–124], Fourier-Mellin transform, wavelet transform [125, 126], and others. And studies of watermarking based on the feature point [127] of original images were proposed.

Matrix decomposition was also considered during the watermark embedding, such as the self-embedding fully blind watermarking algorithm based on QR matrix decomposition [128, 129], and the blind digital watermarking algorithm of QR decomposition of the gray-scale image DWT-FRFT transformation [130], so that image watermarking has good robustness and invisibility.

Literature [131] proposed the quaternion polar harmonic transform watermarking algorithm and realized the three-channel integral watermarking of the color image.

The watermark embedding algorithm based on the DWT transform and the DQFT [132] transform with QR matrix decomposition [133, 134] is proposed, which adds better robustness of the transform domain, embedding the watermark by processing the three channels of color images as a whole.

Related Theories

In the scaling and translation basis function i denotes the direction H,V,D. The wavelet transform of the image f(x,y) is stated as follows, and in the inverse discrete wavelet transform quaternions are an effective representation of a high-dimensional space.

Because of the constraints of the relationship between the three imaginary parts of the quaternion i, j, k, the two-dimensional quaternion Fourier transform can be divided into several types, including the right-sided quaternion Fourier transform, the left-sided quaternion Fourier transform and the two-sided quaternion Fourier transform.

For a color image of $M \times N$ size, $f_R(x, y)$, $f_G(x, y)$, $f_B(x, y)$ can be used to represent the pixel values of the three channels, $1 \leq x \leq M, 1 \leq y \leq N$. Then the elements in the $M \times N$ quaternion matrix can be expressed as a quaternion matrix [125] with 0 as a real part, $f(x, y) = f_R(x, y)i + f_G(x, y)j + f_B(x, y)k$, because of which all channels of the color image are taken into account as a whole.

Watermarking Algorithm

The watermark is embedded in the $r_{1,4}$ element in the first row and fourth column of the R matrix obtained by QR decomposition. Different modification ranges T_1 and T_2 according to the watermark information w are selected. The possible results C_1 and C_2 of the modification based on T_1 and T_2 are determined as below, in which $k = \text{floor}(\text{ceil}(r_{1,4}/t)/2)$, floor(x) is the largest integer not greater than x, ceil(x) is the smallest integer not less than x, and t is the quantization step size, the value of which can be determined through experiments. $r_{1,4}^*$ is calculated after embedding the watermark. Instead of $r_{1,4}$, $r_{1,4}^*$ is used to perform inverse QR decomposition to obtain the 4 × 4 matrix A* with watermarked information. Inverse-QDFT and Inverse-DWT are done to obtain a watermarked color image.

Experiment and Analysis

The correct decoding rate (CDR) is used to evaluate the robustness of the image watermark under the attack of noise, JPEG compression, cropping, and median filtering.

The watermark algorithm in this paper has good robustness against image compression attacks.

Noise is a common attack for image watermarking.

Gaussian noise and salt-and-pepper noise with different variances to attack the image is used to test the correct decoding rate CDR of the watermark.

The experimental results show that the watermarking algorithm in this paper is less robust to Gaussian noise attacks, comparing with JPEG attacks.

Median filtering of different sizes of windows is used to perform filtering on watermarked images to test the robustness against filtering attacks.

Conclusion

The proposed watermarking algorithm of a DWT-DQFT transformation combined with QR decomposition realizes the embedding of watermark information in the three color channels of color images, taking into account the correlation of the three channels of color images.

The watermarking algorithm has good invisibility in SSIM indicators.

The watermark extraction process does not refer to the original color image and the original watermark image, that is the blind extraction of the watermark.

Acknowledgement

A machine generated summary based on the work of Qin, Liangcheng; Ma, Ling; Fu, Xiongjun

2021 in Lecture Notes in Computer Science

5.2 A Robust Blind Watermarking Scheme for Color Images Using Quaternion Fourier Transform

DOI: https://doi.org/10.1007/978-3-030-57881-7_55

Abstract-Summary

Image watermarking is an important method for copyright protection.

Considering that color images dominate the real-world image data, designing copyright protection techniques specially for color images is very meaningful.

We propose a robust blind watermarking scheme for color images based on the quaternion Fourier transform (QFT).

After converting the color image data into QFT coefficients, we embed the watermark message into the four channels in the QFT domain, to improve the embedding capacity and robustness.

Introduction

In [135], the QFT was used to design robust color image watermarking algorithms.

Chen et al. [136] proposed a color image watermarking scheme by combining the QFT and a least squares support vector machine, in order to enhance the robustness against geometrical distortions.

In these existing color watermarking schemes, the watermark messages are embedded into the real channel of the QFT transform domain.

Designing a color image watermarking scheme that can fully use the other three imaginary channels of the QFT transform domain is worthy of study.

We propose a color image watermarking scheme based on the QFT, which has the blind extraction property and is robust against some common watermarking attacks.

Our scheme produces higher perpetual quality watermarked images than the related work [137] that performs embedding only on the real channel of the QFT transform domain.

Preliminaries

It can be used in computer graphics, computer vision, signal/image processing, etc. A quaternion Q has four components, where a is the scalar coefficient of the real channel, and b, c, and d are scalar coefficients of the imaginary channels.

Ell et al. [138] proposed three types of QFTs named left-side QFT, right-side QFT, and both-side QFT.

We adopt the left-side QFT that is widely used in image processing.

Proposed Scheme

We detail our watermark embedding algorithm.

The detailed watermark extraction procedure is described.

Experimental Results

We conduct experiments to study the robustness of the proposed scheme under different types of watermarking attacks.

To further investigate the robust performance of the proposed scheme, we conduct more experiments to compare our method with the schemes with other signal transformation, i.e., DCT and DWT.

We can see that when the cropping occurs at the center or the right bottom corner, the PSNR-BER curves of our scheme are all under the curves of the DCT and WHT, which indicates that our scheme has a higher robust performance than the compared schemes in this case.

According to our experimental results on the robustness to noise attack, cropping attack, and JPEG 2000 compression attack, we can see that the proposed watermarking scheme with QFT has satisfactory robust performance, and is more robust resisting to most of the test watermarking attacks than the schemes with DCT and WHT.

Conclusion
We have proposed a blind robust color image watermarking scheme based on the QFT.

The proposed color watermarking scheme can be applied in nature image copyright protection, source tracking, etc. In the future, our research will focus on the following aspects.

We will combine QFT with DWT, SVD and even deep learning methods, to design a robust color image watermarking scheme against more watermarking attacks, such as rotation attacks and geometrical distortions.

Acknowledgement
A machine generated summary based on the work of Liang, Renjie; Zheng, Peijia; Fang, Yanmei; Song, Tingting
2020 in Lecture Notes in Computer Science

5.3 *New Watermarking Algorithm Utilizing Quaternion Fourier Transform with Advanced Scrambling and Secure Encryption*

DOI: https://doi.org/10.1007/s11042-020-10257-1

Abstract-Summary
Embedding a watermark is done in grayscale images, mainly due to the fact that grayscale images are easier to process than color images, and grayscale images only contain brightness information and color-free information, in which an embedded watermark will not produce new color components.

To improve the security of the watermark information and the ability to embed the location and improve the security of the algorithm against a variety of attacks, this paper proposes an algorithm based on the quaternion Fourier transform (QFT) with chaotic encryption and Arnold scrambling.

We utilized quaternions (which is a subalgebra of GA) and effectively completed color image processing by utilizing the Fourier transform.

After utilizing the QFT, each component was made more secure by scrambling the pixels of the watermark and performing encryption utilizing chaotic sequencing.

Introduction
Despite the different approaches utilized for embedding a watermark, watermarking algorithms still have the following problems for the spatial domain and the quaternion frequency domain: It is not known how to effectively utilize the QFT through other techniques (such as Arnold scrambling and chaotic mapping) to improve the watermark quality and the correlation between the original image and the watermarked image.

The typical algorithms are: Single-channel method: According to the human eye's characteristic of being insensitive to changes in the blue component in the RGB color model, by modifying the blue component value of the color image or transforming it to the transform domain, embed watermark information in the transform domain; [139, 45, 140] or convert the color image from the RGB color model to the YUV or YCbCr color model, and then embed a watermark in the brightness component that is not sensitive to the human eye.

This paper proposes a color image watermarking algorithm based on quaternion theory.

The frequency spectrum is obtained by performing the quaternion Fourier transform (QFT) on the color host image and the watermark image, and the watermark is scrambled.

Background and Working Principles

The quaternion representation of color images is commonly utilized in the literature [57, 141, 142, 139, 45, 140, 143].

Let the three imaginary components of the quaternion represent the three primary color components of red (R), green (G) and blue (B).

The real number is 0, so: The three primary colors in nature are the three coded colors: red (R), green (G), and blue (B).

In the case of color images, the three primary colors of the RGB color model are the color of each pixel, and therefore the three values for R, G and B must be stored in each matrix unit.

Quaternion-based representative color image processing processes its quaternion matrix.

The quaternion approach will better represent the quality of color pictures, as it presents a new path for theoretical advancement or practical implementation, as opposed to the conventional subchannel or grayscale image conversion process.

Every time an Arnold transformation is performed on an image, it is equivalent to a scrambling of the image.

Literature Review

The Directionlet transformation can not only fully capture the detailed information of the color image, but also perform different times of transformation along any two directions, so that the watermark can better resist rotation attacks; in addition, the invariance of the SVD transformation makes the watermarking algorithm robust.

Li and others [144] developed a novel algorithm for handling color images utilizing the quaternion Hadamard transform (QHT) with Schur decomposition.

Hosny and others [145] utilize the quaternion Legendre-Fourier moments (QLFMs) (logistic mapping) to solve the color image watermarking issue by computing the radial and angular kernels over circular pixels utilizing analytical integration.

This approach utilizes quaternions to process correlated image changes to make the watermarking approach more secure against any attacks; however, dual watermarking can be time-consuming.

Study Design and Proposed Algorithm

(1) If it is not a square matrix, convert it into a square after filling null values to obtain w(x, y), which becomes a w(u′, v′) transformation. (2) Apply to f (m, n) the

QFT to obtain the domain graph F(u, v). (3) In this step, find the position of the block where the watermark needs to be embedded by utilizing the frequency shift and replace the high-frequency information of F(u, v) after embedding the watermark to obtain $F_0(u, v)$. (4) Apply the inverse QFT to $F_0(u, v)$ to obtain the image that contains the embedded watermark and record it as $f_0(m, n)$.

The steps of watermark extraction are the inverse of the steps of watermark embedding. (1) Appy the QFT to the image obtained from the previous steps of watermark embedding $f_0(m, n)$ to obtain $F_0(u, v)$. (2) Extract $G(u', v')$ from the high-frequency information in $F_0(u, v)$.

Results and Discussion

The quality evaluation of digital watermarking mainly includes the following two aspects: the subjective or objective quantitative evaluation of the image caused by the embedded watermark and the evaluation of the robustness of the watermark.

The PSNR value is calculated by taking the average of each color component R, G and B. The above indicators are all objective image quality evaluation methods based on full-pixel distortion statistics.

The NC value of all the watermarks of each image is over 90%, while the SSIM value is close to 100%.

It can be seen that compared with other color image watermarking algorithms, our algorithm is approaching the NC value, which shows that our algorithm is more robust than other algorithms.

The difference between the original image and the watermarked image is found in the following by NPCR: where D(i,j) can be calculated as follows: UACI measures the intensity of the original and watermarked images as follows: NPCR and UACI ideal values are 99.6094% and 33.46%, respectively for the gray images [146].

Conclusion

Not only does the color image digital watermarking technology based on blind extraction have strong application value and development prospects, but also the methods and techniques in the research process provide an effective reference for other related technologies.

Although many researchers at home and abroad have conducted fruitful work in this field and obtained many meaningful research and application results, the amount of information contained in the digital color image itself is very large.

Most of the algorithms focus on non-blind extraction, which makes technical research one of the difficulties in the field of digital watermarking.

This paper utilized a QFT with Arnold scrambling and chaotic sequencing to improve the security of a watermark in a color image.

Acknowledgement

A machine generated summary based on the work of Bhatti, Uzair Aslam; Yuan, Linwang; Yu, Zhaoyuan; Li, JingBing; Nawaz, Saqib Ali; Mehmood, Anum; Zhang, Kun

2021 in Multimedia Tools and Applications

5.4 *Quaternion Discrete Fractional Random Transform for Color Image Adaptive Watermarking*

DOI: https://doi.org/10.1007/s11042-017-5511-2

Abstract-Summary
This paper first defines quaternion DFRNT (QDFRNT) which generalizes DFRNT to efficiently process quaternion signals, and then applies QDFRNT to color image adaptive watermarking.

For the QDFRNT, this paper also derives the relationship between QDFRNT of a quaternion signal and DFRNT of four components for this signal to efficiently compute QDFRNT.

For the color image adaptive watermarking based on QDFRNT and SVM, in order to efficiently utilize the color information in the adaptive process, this paper also exploits the human vision system's (HVS) masking properties of texture, edge and color tone directly from the color host image to adaptively adjust the watermark strength for each block.

Experimental results show that: (a) the proposed efficient computation method takes only half the computational time of the direct method; (b) the comparison of five color models (RGB, YUV, YIQ, CIEL*a*b* and YCbCr) shows that the proposed QDFRNT-based watermarking scheme using the YCbCr color model has the overall best performance and can achieve a good balance between invisibility and robustness to the Checkmark attacks; (c) The proposed scheme is superior to the existing schemes respectively using DCT, DFRNT, discrete quaternion Fourier transform (DQFT), discrete fractional quaternion Fourier transform (DFrQFT), and quaternion radial moments (QRMs).

Extended:

As future work, we will improve the scheme to efficiently process high resolution images [147, 148, 149–150].

Introduction
This paper defines quaternion discrete fractional random transform (QDFRNT) to generalize DFRNT to process quaternion signals effectively and then applies QDFRNT to color image adaptive watermarking.

Jiang and others [151] extracted the masking properties of texture, edge and illumination from the graying version of a color host image and then embedded the watermark in the discrete quaternion Fourier transform (DQFT) domain using the adaptive watermark strength.

The main contributions of this paper are as follows, (a) By using the quaternion algebra, we define QDFRNT which generalizes DFRNT to efficiently process quaternion signals in a holistic manner, especially for color image signals. (b) By deriving the relationship between QDFRNT of a quaternion signal and DFRNT of four components for this signal, we propose an efficient computation method for QDFRNT. (c) As one of the applications of QDFRNT, we propose a blind adaptive watermarking scheme for color images based on QDFRNT and SVM.

Some Preliminaries

Quaternions are a generalization of complex numbers - a quaternion has one real part and three imaginary parts given by [152] where a, b, c, d $\in$ R, and i, j, k are three imaginary units. If the real part a = 0, q is called a pure quaternion.

The conjugate and modulus of a quaternion are respectively defined. Let f(x, y) be an RGB image function with the QR, then each pixel can be represented as a pure quaternion where $f_R(x, y)$, $f_G(x, y)$ and $f_B(x, y)$ are respectively the red, green, and blue channels of the pixel (x, y).

Quaternion Discrete Fractional Random Transform (QDFRNT) and Its Efficient Computation

For a 1-D quaternion signal $x_q = x_r + x_i\, i + x_j\, j + x_k\, k$ of size N × 1, its left-side αth-order 1-D QDFRNT is defined. Here the quaternion version of the kernel transform matrix $R^{\alpha,\mu}$ is given with the same V as DFRNT and the different $D^{\alpha,\mu}$. Basically, μ can be defined as a linear combination of i, j, and k such as: μ = a i + b j + c k, a, b, c $\in$ R, |μ| = 1.

For the non-pure quaternion signal, the computational complexity is $32N^3$ multiplications and $(32N^3–8N^2)$ additions for the direct method, while it is $(16N^3 + 24\,N^2)$ multiplications and $(16N^3 + 8\,N^2)$ additions for the proposed method.

For the pure quaternion signal (color image), since the multiplication of an element in the non-pure quaternion matrix $R^{\alpha,\mu}$ and an element in the pure quaternion matrix y_q requires only 12 real number multiplications and 8 additions, the computational complexity is reduced to $24N^3$ multiplications and $(24N^3–8N^2)$ additions for the direct method, while it is $(12N^3 + 18\,N^2)$ multiplications and $(12N^3 + 4\,N^2)$ additions for the proposed method.

Blind Color Image Adaptive Watermarking Scheme Based on QDFRNT and SVM

The proposed scheme improves the drawbacks of the existing color adaptive watermarking schemes: some of them do not efficiently utilize the color information in the adaptive process, or do not consider the holistic property of the channels of the color host image.

We extract the masking properties of texture, edge and color tone directly from color host images to fully use the color information and consider the quaternion-based method to process the color host images in a holistic way.

They proposed a method to compute the color tone as follows: (a) Convert the color image in a RGB model into a CIEL*a*b* model. (b) Normalize the components a* and b*, achieving $f_{an}(x, y)$ and $f_{bn}(x, y)$. (c) Calculate the color tone by where σ is an adjustable parameter (σ = 0.25 in [153] was also considered in this paper).

Experimental Results and Analysis

Taking the results of the DFRNT-based scheme shown in previous tests as a benchmark, the watermarked images of the other five compared schemes were re-generated with similar SSIM values: the DQFT-based one, DFrQFT-based one, and QDFRNT-based one by adjusting the basis watermark strength Δ_0, while the

DCT-based one and the QRMs-based one by multiplying a factor λ to the watermark strength separately used in Refs. [154] and [155].

The main reason is due to the use of quaternion-based idea and the HVS masking properties directly from color host images; (c) for the other five compared schemes, the QRMs-based one performs best, while the DQFT-based one and the DFrQFT-based one, some of whose extracted watermarks are difficult to be recognized, have not obtained satisfactory results.

Conclusion
The QDFRNT has been proposed to generalize the DFRNT to quaternion signal processing.

Theoretical analysis shows that the QDFRNT of a quaternion signal can be efficiently computed through DFRNT of each component of this quaternion signal.

As for the application of the QDFRNT in color image adaptive watermarking, the proposed QDFRNT-based scheme outperforms some other existing schemes for the following reasons: (i) it considers the masking properties directly on the color host image instead of its graying version; (ii) it uses the quaternion-based approach to process three channels of color image holistically instead of treating three channels independently.

The proposed watermarking scheme in this paper mainly processes images with ordinary sizes.

Acknowledgement
A machine generated summary based on the work of Chen, Beijing; Zhou, Chunfei; Jeon, Byeungwoo; Zheng, Yuhui; Wang, Jinwei
2017 in Multimedia Tools and Applications

6 Saliency

Machine Generated Keywords saliency, model, spectral, detection, patch, saliency map, model base, human, map, detection method, state-of-the-art, amplitude, visual, spectrum, base

6.1 Bottom-Up Saliency Detection Model Based on Amplitude Spectrum

DOI: https://doi.org/10.1007/978-3-642-17832-0_35

Original Abstract In this paper, we propose a saliency detection model based on the amplitude spectrum. The proposed model first divides the input image into small patches, and then uses the amplitude spectrum of the Quaternion Fourier Transform (QFT) to represent the color, intensity and orientation distributions for each patch.

The saliency for each patch is determined by two factors: the difference between amplitude spectrums of the patch and its neighbor patches and the Euclidian distance of the associated patches. The novel saliency measure for image patches by using the amplitude spectrum of the QFT proves promising, as the experimental results show that this saliency detection model performs better than the relevant existing models.

Acknowledgement
A machine generated summary based on the work of Fang, Yuming; Lin, Weisi; Lee, Bu-Sung; Lau, Chiew Tong; Lin, Chia-Wen
2011 in Lecture Notes in Computer Science

6.2 Saliency Detection for Stereoscopic 3D Images in the Quaternion Frequency Domain

DOI: https://doi.org/10.1007/s13319-018-0169-8

Abstract-Summary
We propose a saliency detection model for S3D images.

The final saliency map of this model is constructed from the local quaternion Fourier transform (QFT) sparse feature and global QFT log-Gabor feature.

The local QFT feature measures the saliency map of an S3D image by analyzing the location of a similar patch.

The results of experiments on two public datasets show that the proposed model outperforms existing computational saliency models for estimating S3D image saliency.

Extended:

The weighting factor γ is tuned on the NCTU3D database via optimizing the AUC.

Introduction
Fang and others [156] proposed an S3D saliency detection method by which all feature maps, including the depth feature map, are extracted from discrete cosine transformation (DCT) coefficients.

Wang and others [157] presented an S3D saliency detection model by which the depth features are extracted using only the depth contrast, and the depth features are fused with existing 2D saliency features.

Among other related studies, Wang and others [158] presented an S3D saliency detection model in which depth information is employed to generate a depth bias and depth features.

A deep-learning-based S3D saliency detection model proposed by Zhang and others [159] employs the AlexNet model to extract color and depth features and it fuses them into the final saliency map.

The saliency maps are generated through feature detection and by comparing low-level features in the quaternion frequency domain.

Proposed Visual Saliency Model for S3D Images

For the local QFT patch low-frequency sparse-representation method, a disparity patch is first constructed into a quaternion matrix along with all lab color channel maps.

In global QFT saliency detection, the quaternion matrix for the whole image is firstly built using "strengthened" gradient maps of the respective lab color channel and disparity maps.

For each quaternion image patch $C_0(m,n)$, we apply the QFT and use the low-frequency coefficient of each imaginary and real part of the quaternion frequency patch to construct a low-frequency patch.

The saliency value is given by: where y' denotes the quaternion image patch, D' represents the quaternion image dictionary, and x' is the sparse coefficient vector.

In the sparse representation method, each image patch low-frequency vector c(m,n) can be represented, D denotes an image quaternion frequency coefficient dictionary.

Experimental Results

The 3D eye-tracking dataset contains 18 S3D images and the corresponding ground-truth eye fixation density maps processed by a Gaussian kernel.

The NCTU3D saliency dataset contains 475 S3D images along with their depth maps.

Evaluation Criteria: Four widely used evaluation criteria were utilized to quantitatively compare the fixation density map and computed saliency map, namely the linear correlation coefficient (CC), Kullback–Leibler divergence (KLDiv), area under the receiver operating characteristic curve (AUC), and Normalized Scanpath Saliency (NSS).

Parameter Setting: In local saliency detection, we set λ to 0.4 through statistical performance analysis during Lasso linear regression.

In computing the saliency map, most metric resize input images and center-surround (C_s) bias are directly applied in different fixed computing sizes for fast results.

Parameter σ_D in global saliency detection is set to 84.

We compared the proposed 3D saliency model with four 2D saliency models and two S3D saliency models, including ITTI [160], GBVS [161], ICL [162], QFT [163], Fang and others' 3D model [156], and Qi and others' 3D model [164].

Conclusion

Feature extraction in the proposed model only utilizes low level features; high level features such as the human face are neglected.

In most cases, high level features also contained many low level saliency features, especially depth features.

Whether for humans or animal, those high level features always leave a clear depth contour, making S3D saliency detection relatively easier.

This is why the overall performance of the proposed model and a large number of existing S3D saliency detection models increase with depth information.

Acknowledgement
A machine generated summary based on the work of Cai, Xingyu; Zhou, Wujie; Cen, Gang; Qiu, Weiwei
2018 in 3D Research

6.3 Quaternion-Based Spectral Saliency Detection for Eye Fixation Prediction

DOI: https://doi.org/10.1007/978-3-642-33709-3_9

Original Abstract In recent years, several authors have reported that spectral saliency detection methods provide state-of-the-art performance in predicting human gaze in images (see, e.g., [1–3]). We systematically integrate and evaluate quaternion DCT- and FFT-based spectral saliency detection [3, 4], weighted quaternion color space components [5], and the use of multiple resolutions [1]. Furthermore, we propose the use of the eigenaxes and eigenangles for spectral saliency models that are based on the quaternion Fourier transform. We demonstrate the outstanding performance on the Bruce-Tsotsos (Toronto), Judd (MIT), and Kootstra- Schomacker eye-tracking data sets.

Acknowledgement
A machine generated summary based on the work of Schauerte, Boris; Stiefelhagen, Rainer
2012 in Lecture Notes in Computer Science

6.4 Biological Plausibility of Spectral Domain Approach for Spatiotemporal Visual Saliency

DOI: https://doi.org/10.1007/978-3-642-02490-0_31

Original Abstract We provide a biological justification for the success of spectral domain models of visual attention and propose a refined spectral domain based spatiotemporal saliency map model including a more biologically plausible method for motion saliency generation. We base our approach on the idea of spectral whitening (SW), and show that this whitening process is an estimation of divisive normalization, a model of lateral surround inhibition. Experimental results reveal that SW is a better performer at predicating eye fixation locations than other state-of-the-art spatial domain models for color images, achieving a 92% consistency with human behavior in urban environments. In addition, the model is simple and fast, capable of generating saliency maps in real-time.

Acknowledgement
A machine generated summary based on the work of Bian, Peng; Zhang, Liming 2009 in Lecture Notes in Computer Science

7 Motion and Video

Machine Generated Keywords video, visual, ability, realize, target, network, analytic, frame, implementation, model, method, phase, analytic signal, method use, adaptive

7.1 Moving Target Tracking Based on Pulse Coupled Neural Network and Optical Flow

DOI: https://doi.org/10.1007/978-3-319-26555-1_3

Abstract-Summary
The video – particularly video with moving camera – is segmented based on the relative motion occurring between moving targets and background.

By using the fusion ability of pulse coupled neural network (PCNN), the target regions and the background regions are fused respectively.

Using PCNN fuses the direction of the optical flow fusing, and extracts moving targets from video especially with moving camera.

Our video attention map is obtained by means of linearly fusing the above features (direction fusion, phase spectrums and magnitude of velocity), which adds weight for each information channel.

Introduction
Video-based target tracking has drawn increasing interest for its many applications [165], such as video surveillance, traffic control, machine intelligence, biological, medical, etc. In this paper, we are committed at designing a simple simulation system of human vision by combining static information and motion information.

Motion information means PCNN fusion based on optical flow.

A pulse generates from outside to inside based on directional difference in the video frame until a high enough difference value happens.

The saliency map is computed by smoothing linear fusion of phase information, magnitude and direction fusion.

Related Work

This section includes PCNN fusion based on optical flow and topological information extraction.

Optical flow and PCNN applied in our model will be briefly introduced before computational process descriptions of PCNN fusion and topological information are given in detail.

PCNN has the fusion feature, this section combined the PCNN fusion characteristics with the quantitative optical flow direction information which has been pretreated, besides, it respectively fused the target and the background which has the same moving characteristic, so as to realize the separation of foreground and background, and then segment the optical flow moving targets, the calculation process is as follows: Optical flow field pre-processing: for each pixel, optical flow cushioned with a background vector against the maximum optical flow direction, and value 1/10 of the magnitude of the maximum optical flow.

Step 3: The binary image computed from the PCNN_filter is the input of a topological channel which expressed connectivity.

Algorithm Structure

The phase spectrum can be obtained by normalizing the Fourier transform of T, RG and BY.

Phase information is obtained from phase spectrum by the inverse Fourier transform.

Saliency map is computed by smoothing linear fusion of phase information p, magnitude |OF| and direction fusion fus.

Experimental Results

The proposed algorithm is implemented on our video database and compared with FT [166], Vibe [167], PQFT [168].

To further illustrate the effectiveness of the proposed algorithm which combines the visual attention model with PCNN and optical flow, we select the commonly used evaluation index F-Measure to compare the proposed model to FT [166], Vibe [167], PQFT [168].

G is ground truth regions, S is saliency regions: This paper further compares the proposed algorithm with FT [166], Vibe [167], PQFT [168] on the evaluation index F-Measure with different samples.

For videos with moving camera or background disturbance (parachute and birdfall), F-Measures of the proposed model are largest among the four models.

Dealing with video pedestrians and video walking, the proposed model is more effective than FT and PQFT.

Conclusion

This paper proposed a moving target tracking algorithm, which combines the visual attention model with a pulse coupled neural network and optical flow, it has better tracking performance compared with traditional algorithms.

Based on the relative motion occurring between moving targets and background, the target regions and the background are fused respectively by using the fusion ability of PCNN.

Experimental results show that the proposed method has higher detection rate and better ability of suppressing background.

Acknowledgement
A machine generated summary based on the work of Ni, Qiling; Wang, Jianchen; Gu, Xiaodong
2015 in Lecture Notes in Computer Science

7.2 *Fast Correction Visual Tracking via Feedback Mechanism*

DOI: https://doi.org/10.1007/978-3-319-23989-7_22

Abstract-Summary
Most online visual trackers focus on the appearance information and inference theory to realize tracking frame by frame.

Results indicate that the changing values of the target state's posterior distributions provide superior information to the connection between tracking result and the ground truth.

We further analyse the spatial appearance information and propose an adaptive feedback tracking method using the Discrete-Quaternion-Fourier-Transform (DQFT).

Taking advantage of the stability of closed-loop control and the efficiency of the DQFT, the proposed tracker can make a distinction between the easy-tracking frames and the hard-tracking frames, and then re-track hard-tracking frames using further temporal information to realize the correction ability.

Introduction
In each frame, the same pattern – an updating appearance model and a corresponding searching strategy – is applied to achieve the tracking objective without considering the diversity in different frames.

Inspired by the facts mentioned above, we propose a novel tracking method using a feedback mechanism to make a distinction between the easy-tracking frames and the hard-tracking frames.

In order to realize feedback in an online tracking procedure, we analyse the tracking results of different popular methods mentioned in [169, 170].

The proposed method focuses on improving the processing capacity to build a robust tracker, and a flexible tracking strategy is applied to handle different challenges of the targets and scenarios.

Then it exploits the inherent connection between tracking result and the ground truth to generate a feedback mechanism, which realizes a distinctive estimation between easy-tracking frames and hard-tracking frames.

Related Works

The famous mean-shift tracking method [171] was proposed to realize tracking by histogram matching, the spatial information is integrated by statistics theory using histograms and then the tracker maximizes the Bhattacharyya distance iteratively to achieve final results.

Adam proposed a fragments-based robust tracking method [172] based on matching the ensemble of patches to improve the description ability of histograms, more spatial information is exploited by these fragments.

The main contributions of the work are summarized as follows: (1) A novel feedback mechanism model based tracker is proposed, which utilizes the confidence scores to make a judgement and re-track potential failed frames. (2) A temporal-spatial appearance information based model is proposed to re-track the hard-tracking frames, which integrates the spacial information and the temporal information together to realize fine-grained robust tracking. (3) Experimental results on over 50 challenging videos indicate that the proposed tracking method outperforms previous state-of-the-art trackers.

Problem Formulation

We apply and extend the normalized cross correlation method to achieve this objective.

The normalized cross correlation filter [173] is one of the most efficient methods in image matching.

Bolme extended the normalized cross correlation filter to a MOSSE (Minimum Output Sum of Squared Error) form [174].

Danelljan extended the learning scheme for the normalized cross correlation tracker to multichannel color features and proposed a low dimensional adaptive extension of color attributes [175].

Zhang proposed a dense spatio-temporal context tracking scheme using the normalized cross correlation filter [176].

Proposed Tracking Algorithm with Feedback Mechanism

According to confidence scores, we exploit a feedback mechanism to reduce tracking failures in recent frames.

Frames between starting point and end point are defined as hard-tracking frames which should be re-tracked via further processing.

After the hard-tracking frames are defined, we re-track them using a three-step backward search strategy.

Experiments

The FCT method is the second efficient algorithm (average 250 FPS) among all the evaluated methods, just behind the highest speed algorithm CSK (average 320 FPS).

We divide video attributes into six categories: background clutter, motion blur, in-plane rotation, fast motion, out-of-view and occlusion.

We evaluate and compare our FCT method on different attributes.

Motion blur, out-of-view and fast motion: these three attributes are caused by external appearance changing, which can be better solved by the FCT method.

Conclusion
A two-stage FCT tracking algorithm is proposed to achieve a coarse-to-fine tracking.

The FCT method is robust to appearance variations introduced by occlusion, illumination changes, pose variations, motion blur, out-of-view and fast motion.

Experiments with state-of-the-art methods on challenging videos demonstrate that the proposed FCT method achieves favorable results in terms of accuracy, robustness, and speed.

Acknowledgement
A machine generated summary based on the work of Xu, Tianyang; Wu, Xiaojun 2015 in Lecture Notes in Computer Science

7.3 *Desert Vehicle Detection Based on Adaptive Visual Attention and Neural Network*

DOI: https://doi.org/10.1007/978-3-642-42042-9_47

Abstract-Summary
This paper proposes a novel desert vehicle detection method using an adaptive visual attention model and pulse coupled neural network.

An adaptive phase spectrum of the quaternion Fourier transform (APQFT) model is proposed to generate and weight information channels of background, intensity, and image colors into a visual saliency map.

Using a pulse coupled neural network (PCNN) detects regions of interests (ROIs) and using a scale-invariant feature transform (SIFT) extracts features of ROIs.

Acknowledgement
A machine generated summary based on the work of Zhang, Jinjian; Gu, Xiaodong 2013 in Lecture Notes in Computer Science

7.4 *Analytic Video (2D + T) Signals Using Clifford–Fourier Transforms in Multiquaternion Grassmann–Hamilton–Clifford Algebras*

DOI: https://doi.org/10.1007/978-3-0348-0603-9_10

Abstract-Summary
An algebraic framework for (2D + t) video analytic signals and a numerical implementation thereof using Clifford biquaternions and Clifford–Fourier transforms.

After a short presentation of multiquaternion Clifford algebras and Clifford–Fourier transforms, a brief pedagogical review of 1D and 2D quaternion analytic signals using right quaternion Fourier transforms is given.

The biquaternion algebraic framework is developed to express Clifford–Fourier transforms and (2D + t) video analytic signals in standard and polar form constituted by a scalar, a pseudoscalar and six phases.

Acknowledgement
A machine generated summary based on the work of Girard, P. R.; Pujol, R.; Clarysse, P.; Marion, A.; Goutte, R.; Delachartre, P.
2013 in Trends in Mathematics

References

1. C. Fernández, A. Galbis, J. Martínez, Multilinear Fourier multipliers related to time-frequency localization. J. Math. Anal. Appl. 398 (1), 113–122 (2013)
2. R.L. Pego, Compactness in L2 and the Fourier transform. Proc. Am. Math. Soc. 95, 252–254 (1985)
3. Moricz, F.: Absolutely convergent Fourier integrals and classical function spaces. Arch. Math. 91(1), 49–62 (2008)
4. Moricz, F.: Absolutely convergent Fourier series and function classes. J. Math. Anal. Appl. 324(2), 1168–1177 (2006)
5. Moricz, F.: Higher order Lipschitz classes of functions and absolutely convergent Fourier series. Acta Math. Hung. 120(4), 355–366 (2008)
6. Volosivets, S.S.: Fourier transforms and generalized Lipschitz classes in uniform metric. J. Math. Anal. Appl. 383, 344–352 (2011)
7. Volosivets, S.S.: Multiple Fourier coefficients and generalized Lipschitz classes in uniform metric. J. Math. Anal. Appl. (2015). https://doi.org/10.1016/j.jmaa.2015.02.011
8. Achak, A., Bouhlal, A., Daher, R., et al.: Titchmarsh's theorem and some remarks concerning the right-sided quaternion Fourier transform. Bol. Soc. Mat. Mex. 26, 599–616 (2020)
9. Bahri, M., Ashino, R.: A variation on uncertainty principle and logarithmic uncertainty principle for continuous quaternion wavelet transforms. Abstr. Appl. Anal. 2017, 3795120 (2017). https://doi.org/10.1155/2017/3795120
10. M. Bahri, E. Hitzer, A. Hayashi, R. Ashino, An uncertainty principle for quaternion Fourier transform. Comput. Math. Appl. 56 (9), 2398–2410 (2008)
11. Hitzer, E.: The quaternion domain Fourier transform and its properties. Adv. Appl. Clifford Algebras 26(3), 969–984 (2016)
12. Hitzer, E.: General two-sided quaternion Fourier transform, convolution and Mustard convolution. Adv. Appl. Clifford Algebras (preprint). https://doi.org/10.1007/s00006-016-0684-8, http://vixra.org/abs/1601.0165
13. Bahri, M., Ashino, R., Vaillancourt, R.: Two-dimensional quaternion wavelet transform. Appl. Math. Comput. 218(1), 10–21 (2011)
14. Ell, T.A., Bihan, N.L., Sangwine, S.J.: Quaternion Fourier Transforms for Signal and Image Processing. Wiley, New York (2014)
15. Hitzer, E.: Quaternion Fourier transform on quaternion fields and generalizations. Adv. Appl. Clifford Algebra 17, 497–517 (2007). https://doi.org/10.1007/s00006-007-0037-8, preprint: http://arxiv.org/abs/1306.1023
16. Hitzer, E.: Directional uncertainty principle for quaternion Fourier transforms, Adv. Appl. Clifford Algebra 20(2), 271–284 (2010). https://doi.org/10.1007/s00006-009-0175-2, preprint: http://arxiv.org/abs/1306.1276
17. Bülow, T.: Hypercomplex spectral signal representations for the processing and analysis of images, Ph.D. Thesis, University of Kiel 9903, 161 pages (1999)

18. Bülow, T., Felsberg, M., Sommer, G.: Non-commutative hypercomplex Fourier transforms of multidimensional signals. In: Sommer, G. (ed.) Geometric Computing with Clifford Algebras. Theor. Found. and Appl. in Comp. Vision and Robotics, pp. 187–207. Springer, Berlin (2001)
19. Antoine, J.P., Murenzi, R.: Two-dimensional directional wavelets and the scale-angle representation. Sig. Process. 52(3), 259–281 (1996)
20. Antoine, J.P., Vandergheynst, P., Murenzi, R.: Two-dimensional directional wavelets in image processing. Int. J. Imag. Syst. Technol. 7(3), 152–165 (1996)
21. Bahri, M.: Quaternion algebra-valued wavelet transform. Appl. Math. Sci. 5(71), 3531–3540 (2011)
22. Hitzer, E., Sangwine, S.J. (eds.): Quaternion and Clifford–Fourier Transforms and Wavelets, Trends in Mathematics, pp. 57-83, Springer, Basel (2013)
23. Chen, L.-P., Kou, K.I., Liu, M.-S.: Pitt's inequality and the uncertainty principle associated with the quaternion Fourier transform. J. Math. Anal. Appl. (2015)
24. Guanlei, X., Xiaotong, W., Xiaogang, X.: Fractional quaternion Fourier transform, convolution and correlation. Signal Process. 88(10), 2511–2517 (2008)
25. Guo, L., Zhu, M., Ge, X.: Reduced biquaternion canonical transform, convolution and correlation. Signal Process. 91(8), 2147–2153 (2011)
26. Kou, K.I., Morais, J.: Asymptotic behaviour of the quaternion linear canonical transform and the Bochner–Minlos theorem. Appl. Math. Comput. 247(15), 675–688 (2014)
27. Yang, Y., Kou, K.I.: Uncertainty principles for hypercomplex signals in the linear canonical transform domains. Signal Process. 95, 67–75 (2014)
28. Kou, K.I., Ou, J.Y., Morais, J.: On uncertainty principle for quaternionic linear canonical transform. Abstr. Appl. Anal. (Hindawi Publishing Corporation) 2013, 14 (2013) (article ID 725952)
29. Bahri, M., Saleh Arif, F.M.: Relation between quaternion Fourier transform and quaternion Wigner–Ville distribution associated with linear canonical transform. J. Appl. Math. 2017 (article ID 3247364)
30. Bracewell, R.: The Fourier Transform and Its Applications, 3rd edn. McGraw-Hill Book Company, New York (2000)
31. Chen, D., Fečkan, M., Wang, J.: On the stability of linear quaternion-valued differential equations. Qual. Theory Dyn. Syst. 21, 1–7 (2022)
32. Chen, D., Fečkan, M., Wang, J.: Hyers-Ulam stability for linear quaternion-valued differential equations with constant coefficient. Rocky Mt. J. Math., (2021), https://projecteuclid.org/journals/rmjm/rocky-mountain-journal-of-mathematics/DownloadAcceptedPapers/210126-Wang.pdf
33. Cheng, D., Kou, K.I., Xia, Y.H.: Floquet theory for quaternion-valued differential equations. Qual. Theory Dyn. Syst. 19, 1–23 (2020)
34. Amrein, W.O., Berthier, A.M.: On support properties of Lp-functions and their Fourier transforms. J. Funct. Anal. 24, 258–267 (1977). https://doi.org/10.1016/0022-1236(77)90056-8
35. Tyr, O., Daher, R.: Benedicks–Amrein–Berthier type theorem and local uncertainty principles in Clifford algebras. Rend. Circ. Mat. Palermo II Ser. (2021). https://doi.org/10.1007/s12215-021-00669-9
36. Ell, T.A.: In: Fourier Transforms for Analysis of Two-Dimensional Linear Time-Invariant Partial Differential Systems, pp. 1830–1841. San Antonio, Texas (1993)
37. Hitzer, E.; Sangwine, S.J.: The orthogonal 2D planes split of quaternions and steerable quaternion Fourier transformations, in: E. Hitzer and S.J. Sangwine (Eds.), Quaternion and Clifford Fourier Transforms and Wavelets, Trends in Mathematics, vol. 27, pp. 15–40. Birkhäuser (2013) https://doi.org/10.1007/978-3-0348-0603-9_2, preprint: http://arxiv.org/abs/1306.2157
38. El Haoui, Y., Fahlaoui, S.: The uncertainty principle for the two-sided quaternion Fourier transform. Mediterr. J. Math. (2017). https://doi.org/10.1007/s00009-017-1024-5

39. El Haoui, Y., Fahlaoui, S.: Miyachi's Theorem for the Quaternion Fourier Transform. Circ. Syst. Sig. Process 39, 2193–2206 (2020). https://doi.org/10.1007/s00034-019-01243-6
40. Y. El Haoui, S. Fahlaoui, Beurling's theorem for the quaternion Fourier transform. J. Pseudo-Differ. Oper. Appl. (2019). https://doi.org/10.1007/s11868-019-00281-7
41. K.M. Hosny, Y.M. Khedr, W.I. Khedr et al., Robust color image hashing using quaternion polar complex exponential transform for image authentication. Circuits Syst. Signal Process. 37, 5441 (2018). https://doi.org/10.1007/s00034-018-0822-8
42. Folland, G.B., Sitaram, A.: The uncertainty principle: a mathematical survey. J. Fourier Anal. Appl. 3(3), 207–238 (1997)
43. Thangavelu, S.: An Introduction to the Uncertainty Principle, Progress in Mathematics, vol. 217. Birkhauser, Boston (2004)
44. Donoho, D.L., Stark, P.B.: Uncertainty principles and signal recovery. SIAM J. Appl. Math. 49(3), 906–931 (1989)
45. S.C. Pei, J.J. Ding, J.H. Chang, Efficient implementation of quaternion Fourier transform, convolution, and correlation by 2-D complex FFT. IEEE Trans. Signal Process. 49 (11), 2783–2797 (2001)
46. Christensen, J. G.: Uncertainty principle. Master's Thesis, Institute for Mathematical science, University of Copenhagen (2003)
47. Hahn, S.L., Snopek, K.M.: Wigner distributions and ambiguity function of 2-D quaternionic and monogenic signals. IEEE Trans. Sigal Process. 53(8), 3111–3128 (2005)
48. Lamouchi, H., Omri, S.: Quantitative uncertainty principles for the short time Fourier transform and the radar ambiguity function. Indian J. Pure Appl. Math. 48(1), 147–161 (2017)
49. Chen, Q., Qian, T.: Sampling theorem and multi-scale spectrum based on non-linear Fourier atoms. Appl. Anal. 88(6), 903–919 (2009)
50. Chen, Q., Wang, Y., Wang, Y.: A sampling theorem for non-bandlimited signals using generalized sinc functions. Comput. Math. Appl. 56(6), 1650–1661 (2008)
51. Liu, Y.L., Kou, K.I., Ho, I.T.: New sampling formulae for non-bandlimited signals associated with linear canonical transform and nonlinear Fourier atoms. Signal Process. 90(3), 933–945 (2010)
52. Cheng, D., Kou, K.I.: Novel sampling formulas associated with quaternionic prolate spheroidal wave functions. Adv. Appl. Clifford Algebras 27(4), 2961–2983 (2017)
53. Cheng, D., Kou, K.I.: Generalized sampling expansions associated with quaternion Fourier transform. Math. Methods Appl. Sci. 41(11), 4021–4032 (2018)
54. Hu, X., Cheng, D., Kou, K.: Sampling formulas for 2D quaternionic signals associated with various quaternion Fourier and linear canonical transforms. Front. Inf. Technol. Electr. Eng. (2021)
55. Xiao-xiao, H., Kou, K.I.: Inversion theorems of quaternion Fourier and linear canonical transforms. Math. Methods Appl. Sci. 40(7), 2421–2440 (2017)
56. Xiang, M., Dees, B.S., Mandic, D.P.: Multiple-model adaptive estimation for 3-D and 4-D signals: a widely linear quaternion approach. IEEE Trans. Neural Netw. Learn. Syst. 30(1), 72–84 (2019)
57. Chen, Y., Xiao, X., Zhou, Y.: Low-rank quaternion approximation for color image processing. IEEE Trans. Image Process. 29, 1426–1439 (2020)
58. Kaur, H., Kumar, M., Sharma, A.K., Singh, H.P.: Performance analysis of different Wavelet families over fading environments for mobile WiMax system. Int. J. Future Gener. Commun. Netw. 8, 87–98 (2015)
59. Davis, J.A., Jedwab, J.: Peak-to-mean power control in OFDM, Golay complementary sequences, and Reed-Muller codes. IEEE Trans. Inf. Theory 45, 2397–2417 (1999)
60. Michailow, N., Mendes, L., Matthe, M., Festag, I., Fettweis, A., Robust, G.: WHT-GFDM for the next generation of wireless networks. IEEE Commun. Lett. 19, 106–109 (2015)
61. Manhas, P., Soni, M.K.: Comparison of OFDM system in terms of BER using different transform and channel coding. Int. J. Eng. Manuf. 1, 28–34 (2016)
62. Labunets, V.G.: Quaternion number–theoretical transform. In: Devices and Methods of Experimental Investigations in Automation, pp. 28–33. Dnepropetrovsk State University Press, Dnepropetrovsk (1981). (In Russian)

63. Sommen, F.: A product and an exponential function in hypercomplex function theory. Appl. Anal. 12, 13–26 (1981)
64. Sommen, F.: Hypercomplex Fourier and Laplace transforms I. Ill. J. Math. 26(2), 332–352 (1982)
65. Labunets-Rundblad, E.: Fast Fourier-Clifford transforms design and application in invariant recognition. Ph.D. thesis, p. 26. Tampere University Technology, Tampere, Finland (2000)
66. Rundblad, E., Labunets, V., Egiazarian, K., Astola, J.: Fast invariant recognition of color images based on Fourier–Clifford number theoretical transform. In: EUROPORTO, Conference on Image and Signal Processing for Remote Sensing YI, pp. 284–292 (2000)
67. Labunets, V.G., Kohk, E.V., Ostheimer, E.: Algebraic models and methods of computer image processing. Part 1. Multiplet models of multichannel images. Comput. Opt. 42(1), 84–96 (2018)
68. Jorswieck, E., Tomasin, S., Sezgin, A.: Broadcasting into the uncertainty: authentication and confidentiality by physical-layer processing. Proc. IEEE 103(10), 1702–1724 (2015)
69. Wyner, A.D.: The wiretap channel. Bell Labs Tech. J. 54(8), 1355–1387 (1975)
70. Renna, F., Laurenti, N., Poor, H.V.: Physical-layer secrecy for OFDM transmissions over fading channels. IEEE Trans. Inf. Forens. Secur. 7(4), 1354–1367 (2012)
71. Chorti, A., Poor, H.V.: Faster than Nyquist interference assisted secret communication for OFDM systems. In: Proceedings of the IEEE Asilomar Conference on Signals, Systems and Computers, pp. 183–187 (2011)
72. Wang, X.: Power and subcarrier allocation for physical-layer security in OFDMA-based broadband wireless networks. IEEE Trans. Inf. Forens. Secur. 6(3), 693–702 (2011)
73. Wang, H.M., Yin, Q., Xia, X.G.: Distributed beamforming for physical-layer security of two-way relay networks. IEEE Trans. Signal Process. 60(7), 3532–3545 (2012)
74. Gupta, M.K., Tiwari, S.: Performance evaluation of conventional and wavelet based OFDM system. Int. J. Electron. Commun. 67(4), 348–354 (2013)
75. Halford, K., Halford, S., Webster, M., Andren, C.: Complementary code keying for rake-based indoor wireless communication. In: Proceedings of IEEE International Symposium on Circuits and Systems, pp. 427–430 (1999)
76. Golay, M.J.E.: Complementary series. IEEE Trans. Inform. Theory 7, 82–87 (1961)
77. Xiao, J., Yu, J., Li, X., Tang, Q., Chen, H., Li, F., Cao, Z., Chen, L.: Hadamard transform combined with companding transform technique for PAPR reduction in an optical direct-detection OFDM system. IEEE J. Opt. Commun. Netw. 4(10), 709–714 (2012)
78. Wilkinson, T.A., Jones, A.E.: Minimization of the peak to mean envelope power ratio of multicarrier transmission schemes by block coding. In: Proceedings of the IEEE 45th Vehicular Technology Conference, pp. 825–829 (1995)
79. Wilkinson, T.A., Jones, A.E.: Combined coding for error control and in creased robustness to system nonlinearities in OFDM. In: Proceedings of the IEEE 46th Vehicular Technology Conference, pp. 904–908 (1996)
80. Fu, Y.X., Li, L.Q.: Paley-Wiener and Boas theorems for the quaternion Fourier transform. Adv. Appl. Clifford Algebr. 23, 837–848 (2013)
81. Plataniotis, K.N., Venetsanopoulos, A.N.: Color Image Processing and Applications. Springer, Berlin (2000)
82. Turkmen, I.: The ANN based detector to remove random-valued impulse noise in images. J. Vis. Commun. Image Represent. 34, 28–36 (2016)
83. Nair, M.S., Shankar, V.: Predictive-based adaptive switching median filter for impulse noise removal using neural network-based noise detector. Signal Image Video Process. 7(6), 1041–1070 (2013)
84. Kaliraj, G., Baskar, S.: An efficient approach for the removal of impulse noise from the corrupted image using neural network based impulse detector. Image Vis. Comput. 28, 458–466 (2010)
85. Liang, S.F., Lu, S.M., Chang, J.Y., Lin, C.T.: A novel two-stage impulse noise removal technique based on neural networks and fuzzy decision. IEEE Trans. Fuzzy Syst. 16(4), 863–873 (2008)

86. Jin, L., Jin, M., Xu, X., Song, E.: Structure-adaptive vector median filter for impulse noise removal in color images. In: IEEE International Conference on Image Processing (ICIP), pp. 690–694 (2017)
87. Zhang, W., Jin, L., Song, E., Xu, X.: Removal of impulse noise in color images based on convolution neural network. Appl. Soft Comput. J. 82, 10558 (2019)
88. Lin, T.-C.: Decision-based filter based on SVM and evidence theory for image noise removal. Neural Comput. Appl. 21(4), 695–703 (2012)
89. Kashyap RL, Khotanzed A (1986) A model based method for rotation invariant texture classification. IEEE Trans Pattern Anal Machine Intell PAMI-8(4):472–481
90. Chantler M, Schmidt M, Petrou M, McGunnigle G (2002) The effect of illuminant rotation on texture filters: Lissajous's ellipses. Proc Eur Conf Comput Vision 3:289–303
91. Vertan C, Boujemaa N (2000) Color texture classification by normalized color space representation. In: 15th International conference on pattern recognition (ICPR'00), vol 3, Barcelona, pp. 35–84
92. Shi L, Funt B (2005) Quaternion color texture, AIC'2005 proceedings of the tenth congress of the international color association, Granada
93. Sangwine S.J.: Fourier transforms of colour images using quaternion, or hypercomplex, numbers. Electron. Lett. 32(21), 1979–1980 (1996)
94. Davis LS, Johns SA, Aggarwal JK (1979) Texture analysis using generalized co-occurrence matrices. IEEE Trans Pattern Anal Machine Intell PAMI-1:251–259
95. Haralick RM, Shanmugam K, Dinstein I (1973) Textural features for image classification. IEEE Trans Systems, Man Cybernet 3:610–621
96. Hu M (1962) Visual pattern recognition by moment invariants. IRE Trans Inform Theory 8:179–187
97. Ketterings QM, Coe R, van Noordwijk M, Palm CA (2001) Reducing uncertainty in the use of allometric biomass equations for predicting above-ground tree biomass in mixed secondary forests. For Ecol Manag 146:199–209
98. Chave J, Condit R, Aguilar S, Hernandez A, Lao S, Perez R (2004) Error propagation and scaling for tropical forest biomass estimates. Philos Trans R Soc Biol Sci 359:409–420
99. Chave J, Rejou-Mechain M, Burquez A, Chidumayo E, Colgan MS, Delitti WBC, Duque A, Eid T, Fearnside PM, Goodman RC, Henry M, Martinez-Yrizar A, Mugasha WA, Muller-Landau HC, Mencuccini M, Nelson BW, Ngomanda A, Nogueira EM, Ortiz-Malavassi E, Pelissier R, Ploton P, Ryan CM, Saldarriaga JG, Vieilledent G (2014) Improved allometric models to estimate the above ground biomass of tropical trees. Glob Change Biol. https://doi.org/10.1111/gcb.12629
100. Basuki TM, Van Laake PE, Skidmore AK, Hussin YA (2009) Allometric equations for estimating the above-ground biomass in tropical lowland Dipterocarp forests. For Ecol Manag 257:1684–1694. https://doi.org/10.1016/j.foreco.2009.01.027
101. Saatchi SS, Harris NL, Brown S, Lefsky M, Mitchard ET, Salas W, Morel A (2011) Benchmark map of forest carbon stocks in tropical regions across three continents. Proc Natl Acad Sci 108:9899–9904
102. Feldpausch TR, Lloyd J, Lewis SL, Brienen RJ, Gloor M, Monteagudo Mendoza A, Lopez-Gonzalez G, Banin L, Abu Salim K, Affum-Baffoe K, Alexiades M (2012) Tree height integrated into pantropical forest biomass estimates. Biogeosciences 9:3381–3403. https://doi.org/10.5194/bg-9-3381-2012
103. Vieilledent G, Vaudry R, Andriamanohisoa SF, Rakotonarivo OS, Randrianasolo HZ, Razafindrabe HN, Rakotoarivony CB, Ebeling J, Rasamoelina M (2012) A universal approach to estimate biomass and carbon stock in tropical forests using generic allometric models. Ecol Appl 22:572–583
104. Fayolle A, Doucet JL, Gillet JF, Bourland N, Lejeune P (2013) Tree allometry in Central Africa: testing the validity of pantropical multi-species allometric equations for estimating biomass and carbon stocks. For Ecol Manag 305:29–37

105. Hunter MO, Keller M, Victoria D, Morton DC (2013) Tree height and tropical forest biomass estimation. Biogeosciences 10:8385–8399
106. Kearsley E, De Haulleville T, Hufkens K, Kidimbu A, Toirambe B, Baert G, Verbeeck H (2013) Conventional tree height–diameter relationships significantly overestimate aboveground carbon stocks in the Central Congo Basin. Nat Commun. https://doi.org/10.1038/ncomms3269
107. Mitchard ET, Saatchi SS, Baccini A, Asner GP, Goetz SJ, Harris NL, Brown S (2013) Uncertainty in the spatial distribution of tropical forest biomass: a comparison of pan-tropical maps. Carbon Balance Manag 8:1–13
108. Ekoungoulou R, Liu X, Loumeto JJ, Ifo SA, Bocko YE, Koula FE, Niu S (2014) Tree allometry in tropical forest of Congo for carbon stocks estimation in aboveground biomass. Open J For 4(05):481
109. Ekoungoulou R, Niu S, Loumeto JJ, Ifo SA, Bocko YE, Mikieleko FEK, Liu X (2015) Evaluating the carbon stock in above-and below-ground biomass in a moist Central African forest. Sci Educ 2:51–59
110. Picard N, Bosela FB, Rossi V (2014) Reducing the error in biomass estimates strongly depends on model selection. Ann For Sci 72:811–823
111. Gaia VL, Qi C, Jeremy AL, David AC, Del Frate Fabio, Leila G, Francesco P, Riccardo V (2014) Aboveground biomass estimation in an African tropical forest with lidar and hyperspectral data. J Photogramm Remote Sens 89:49–58
112. Re DS, Engel VL, Sousa OLM, Blanco JLA (2015) Tree allometric equations in mixed forest plantations for the restoration of seasonal semi deciduous forest. CERNE 21:133–140. https://doi.org/10.1590/01047760201521011452
113. Couteron P (2002) Quantifying change in patterned semi-arid vegetation by Fourier analysis of digitized aerial photographs. Remote Sens 23:3407–3425
114. Couteron P, Raphael P, Eric A, Domonique P (2005) Predicting tropical forest stand structure parameters from Fourier transform of very high-resolution remotely sensed canopy images. J Appl Ecol 42:1121–1128
115. Sangwine SJ, Ell TA (1999) Hypercomplex auto-and cross-correlation of color images. In: IEEE international conference on image processing (ICIP'99), Kobe, Japan, pp. 319–322
116. Ell T.A., Sangwine S.J.: Hypercomplex Fourier transforms of color images. IEEE Trans. Image Process. 16(1), 22–35 (2007)
117. Moxey C.E., Sangwine S.J., Ell T.A.: Hypercomplex correlation techniques for vector images. IEEE Trans. Signal Process. 51, 1941–1953 (2003)
118. Shi L, Funt B (2005) Quaternion colour texture. In: Proceedings 10th congress of the international color association, Granada
119. Mugglestone MA, Renshaw E (1996) A practical guide to the spectral analysis of spatial point processes. Comput Stat Data Anal 21:43–65
120. Tapamo H, Mfopou A, Ngonmang B, Couteron P, Monga O (2014) Linear versus non-linear methods: a comparative study for forest above ground biomass estimation from texture analysis of satellite image. ARIMA 18:114–131
121. Guo, H., Du, Y., Xu, Q.: Quantum image watermarking algorithm based on blocked spatial domain. Chin. J. Quant. Electron. 35(5), 527–532 (2018)
122. Zhang, X., Xiao, Y., Zhao, Z.: Self-embedding fragile watermarking based on DCT and fast fractal coding. Multimedia Tools Appl. 74(15), 5767–5786 (2014)
123. Fu, J., Chen, D., Xu, D., Mao, J.: A watermarking algorithm for image content authentication in double-compression environment. Scientia Sinica Informationis 49(4), 464–485 (2019)
124. Zhang, N.N., Yu, L., Yang, X.F.: Research of digital image watermarking robustness algorithm based on DCT[C]. Prog. Appl. Sci. Eng. Technol. Source Adv. Mater. Res. 926–930, 3171–3174 (2014)
125. Ma, L., Zhang, X.: Characteristics of color images with watermark based on the relationship between non-void subspaces of inner space. Chinese J. Comput. 40(5), 1204–1217 (2017)

126. Hai, F., Quan, Z., Kaijia, L.: Robust watermarking scheme for multispectral images using discrete wavelet transform and tucker decomposition. J. Comput. 8(11), 2844–2850 (2013)
127. Liu, Q., Zhang, L., Zhang, Y., et al.: Geometrically synchronous watermarking algorithm based on the corner feature. J. Commun. 32(4), 25–31 (2011)
128. Han, S., Zhang, H.: Self-embedding perfectly blind watermarking algorithm based on QR decomposition for color images. J. Graphics 36(03), 345–351 (2015)
129. Chen, Y.: Structure-preserving QR algorithm of general quaternion eigenvalue problem with application to color watermarking. Jiangsu Normal University (2018)
130. Wu, Q., Peng, Y.: A blind digital watermarking algorithm based on DWT-FRFT transform and QR decomposition. Electron. Sci. Tech. 31(10), 53–55 (2018)
131. Liu, Y., Zhang, S., Yang, J.: Color image watermark decoder by modeling quaternion polar harmonic transform with BKF distribution. Signal Process.: Image Commun. 88, 115946 (2020)
132. Guo, J., Ma, Y.: Color image digital watermarking algorithm based on quaternion Fourier transform. Packag. Eng. 38(3), 155–159 (2017)
133. Rasti, P., Anbarjafari, G., Demirel, H.: Colour image watermarking based on wavelet and QR decomposition. In: 2017, 25th Signal Processing and Communications Applications Conference (SIU), Antalya, pp. 1–4 (2017). https://doi.org/10.1109/SIU.7960259
134. Liu, Y., Wang, J., Hu, H., et al.: Robust blind digital watermarking scheme based on contourlet transform and QR decomposition. J. Optoelectr. Laser 27(3), 317–324 (2016)
135. Wang, X., Wang, C., Yang, H., Niu, P.: A robust blind color image watermarking in quaternion Fourier transform domain. J. Syst. Softw. 86(2), 255–277 (2013)
136. Chen, B., Coatrieux, G., Gang, C., Sun, X., Coatrieux, J.L., Shu, H.: Full 4-D quaternion discrete Fourier transform based watermarking for color images. Digit. Sig. Process. 28(25), 106–119 (2014)
137. Bas, P., Bihan, N.L., Chassery, J.: Color image watermarking using quaternion Fourier transform, vol. 3, pp. 521–524 (2003)
138. Ell, T.A., Sangwine, S.J.: Decomposition of 2D hypercomplex Fourier transforms into pairs of complex Fourier transforms. In: 2000 European Signal Processing Conference, pp. 1–4 (2000)
139. Jin L, Li D (2007) An efficient color-impulse detector and its application to color images. IEEE Signal Processing Letters 14(6):397–400
140. Sangwine S.J.: Color image edge detector based on quaternion convolution. Electron. Lett. 34, 969–971 (1998)
141. Ell TA, Le Bihan N, Sangwine SJ (2014) Quaternion Fourier transforms for signal and image processing. John Wiley & Sons
142. Jiang S (2008) "The theory and application research of color image processing based on hyper-complex", Ph.D. dissertation. Fudan University, Shanghai
143. Wang XY, Liu YN, Han MM, Yang HY (2016) Local quaternion PHT based robust color image watermarking algorithm. J Vis Commun Image Represent 38:678–694
144. Li J, Yu C, Gupta BB, Ren X (2018) Color image watermarking scheme based on quaternion Hadamard transform and Schur decomposition. Multimed Tools Appl 77:4545–4561
145. Hosny KM, Darwish MM (2019) Invariant color images representation using accurate quaternion Legendre–Fourier moments. Pattern Anal Applic 22(3):1105–1122
146. Nestor T, De Dieu NJ, Jacques K, Yves EJ, Iliyasu AM, El-Latif A, Ahmed A (2020) A multidimensional hyperjerk oscillator: dynamics analysis, analogue and embedded systems implementation, and its application as a cryptosystem. Sensors 20(1):83
147. Xie X, Livermore C (2016) A pivot-hinged, multilayer SU-8 micro motion amplifier assembled by a self-aligned approach. In: Proc. 2016 I.E. 29th Int. Conf. Micro Electro Mechanical Systems (MEMS), Shanghai, pp 75–78
148. Xie X, Livermore C (2017) Passively self-aligned assembly of compact barrel hinges for high-performance, out-of-plane mems actuators. In Proc. 2017 I.E. 30th Int. Conf. Micro Electro Mechanical Systems (MEMS), Las Vegas, pp 813–816

149. Xie X, Zaitsev Y, Velásquez-García LF, Teller S, Livermore C (2014) Scalable, MEMS-enabled, vibrational tactile actuators for high resolution tactile displays. J Micromech Microeng 24(12):125014
150. Xie X, Zaitsev Y, Velásquez-García LF, Teller S, Livermore C (2014) Compact, scalable, high-resolution, MEMS-enabled tactile displays. In: Proc. 30th Solid-State Sensors, Actuators, and Microsystems Workshop, Hilton Head Island, South Carolina, pp 127–130
151. Jiang SH, Zhang JQ, Hu B (2009) An adaptive watermarking algorithm in the hypercomplex space of a color image. Acta Electron Sin 37(8):1773–1778 (in Chinese)
152. Hamilton WR (1866) Elements of quaternions. Longmans, Green, & Company, London
153. Tsai HH, Sun DW (2007) Color image watermark extraction based on support vector machines. Inf Sci 177(2):550–569
154. Kalra GS, Talwar R, Sadawarti H (2015) Adaptive digital image watermarking for color images in frequency domain. Multimed Tools Appl 74(17):6849–6869
155. Tsougenis ED, Papakostas GA, Koulouriotis DE, Karakasis EG (2014) Adaptive color image watermarking by the use of quaternion image moments. Expert Syst Appl 41(14):6408–6418
156. Fang, Y., Wang, J., Narwaria, M., et al. (2014). Saliency detection for stereoscopic images. IEEE Transactions on Image Processing, 23(6), 2625–2636.
157. Wang, J., Da Silva, M. P., Le Callet, P., et al. (2013). Computational model of stereoscopic 3D visual saliency. IEEE Transactions on Image Processing, 22(6), 2151–2165.
158. Wang, A., & Wang, M. (2017). RGB-D salient object detection via minimum barrier distance transform and saliency fusion. IEEE Signal Processing Letters, 24(5), 663–667.
159. Zhang, Q., Wang, X., Jiang, J. et al. (2016). Deep learning features inspired saliency detection of 3D images. In Pacific rim conference on multimedia, (pp. 580–589).
160. Itti, L., Koch, C., Niebur, E.: A model of saliency-based visual attention for rapid scene analysis. IEEE Transactions on Pattern Analysis and Machine Intelligence 20(11), 1254–1259 (1998)
161. Harel, J., Koch, C., Perona, P. (2006). Graph-based visual saliency. In Proceedings of advances in neural information processing systems (NIPS), (pp. 545–552).
162. Li, Y., Zhou, Y., & Xu, L. (2009) Incremental sparse saliency detection. In 6th IEEE international conference on image processing (ICIP), (pp. 3093–3096).
163. Fang, Y., Lin, W., Lee, B. S., et al. (2012). Bottom-up saliency detection model based on human visual sensitivity and amplitude spectrum. IEEE Transactions on Multimedia, 14(1), 187–198.
164. Qi, F., Zhao, D., Liu, S., et al. (2017). 3D visual saliency detection model with generated disparity map. Multimedia Tools and Applications, 76(2), 3087–3103.
165. Kim, W., Kim, C.: Spatiotemporal saliency detection using textural contrast and its applications. IEEE Trans. Circuits Syst. Video Technol. 24, 646–659 (2014)
166. Achanta, R., Hemami, S., Estrada, F., Susstrunk, S.: Frequency-tuned salient region detection. In: IEEE CVPR, pp. 1597–1604 (2009)
167. Barnich, Olivier, Van Droogenbroeck, Marc: Vibe: a universal background subtraction algorithm for video sequences. IEEE Trans. Image Process. 20(6), 1709–1724 (2011)
168. Guo, C.L., Ma, Q., Zhang, L.M.: Spatio-temporal saliency detection using phase spectrum of quaternion Fourier transform. In: IEEE Conference on Computer Vision and Pattern Recognition, pp, 1–8(2008)
169. Wu, Y., Lim, J., Yang, M.-H.: Online object tracking: A benchmark. In: IEEE CVPR, pp. 2411–2418. IEEE Press, Portland (2013)
170. Smeulders, A.W.M., Chu, D.M., Cucchiara, R., Calderara, S., Dehghan, A., Shah, M.: Visual tracking: an experimental survey. IEEE Trans. Pattern Anal. Mach. Intell. 36(7), 1442–1468 (2014)
171. Comaniciu, D., Ramesh, V., Meer, P.: Real-time tracking of non-rigid objects using mean shift. In: IEEE CVPR (2000)
172. Adam, A., Rivlin, E., Shimshoni, I.: Robust fragments-based tracking using the integral histogram. In: IEEE CVPR (2006)

173. Briechle, K., Hanebeck, U.D.: Template matching using fast normalized cross correlation. In: SPIE, vol. 4387, pp. 95–102 (2001)
174. Bolme, D.S., Beveridge, J.R., Draper, B.A., Lui, Y.M.: Visual object tracking using adaptive correlation filters. In: IEEE CVPR (2010)
175. Danelljan, M., Khan, F.S., Felsberg, M., Weijer, J.: Adaptive color attributes for real-time visual tracking. In: IEEE CVPR, pp. 1090–1097 (2014)
176. Zhang, K., Zhang, L., Liu, Q., Zhang, D., Yang, M.-H.: Fast visual tracking via dense spatio-temporal context learning. In: Fleet, D., Pajdla, T., Schiele, B., Tuytelaars, T. (eds.) ECCV 2014, Part V. LNCS, vol. 8693, pp. 127–141. Springer, Heidelberg (2014)

Quaternion Wavelet Transform (QWT)

Eckhard Hitzer

Chapter Introduction Fourier transforms lack localization, this is remedied using localized wavelets of various shapes that can be shifted, scaled and rotated. Naturally this has also been applied for obtaining *quaternionic wavelet transforms*.

Section 1.1 provides retrospective, theory and new applications of the QWT. It covers both discrete and continuous quaternionic wavelets, compactly supported and orthogonal wavelets with symmetry and the associated uncertainty.

Section 1.2 shows how the QWT has been further developed in mathematics toward more general linear canonical wavelets, for functions and function distributions, for quaternionic Dini Lipschitz functions and in the study of quaternionic K-functionals and smoothness moduli.

Section 2 surveys the application of the QWT in the field of image processing. Section 2.1 focusses on multi-resolution image analysis, quaternionic atomic functions, colorization and the establishment of an image sharpness metric. Section 2.2 treats image denoising with QWT and fractional QWT, and information-based scale saliency.

Section 3 introduces a range of color image watermarking schemes and forensic feature analysis for the distinction of photographs from computer graphics.

Finally, Section 4 surveys multi-modal medical image fusion with the QWT, skin lesion classification, echocardiography temporal enhancement and visual object tracking and identification.

Machine Generated Keywords wavelet, QWT, wavelet transform, quaternion wavelet, filter, coefficient, CQWT, fusion, phase, subband, scheme, method, watermarking, feature, algorithm

E. Hitzer (✉)
International Christian University, Mitaka, Tokyo, Japan
e-mail: hitzer@icu.ac.jp

E. Hitzer (ed.), *Quaternionic Integral Transforms*, Trends in Mathematics,
https://doi.org/10.1007/978-3-031-28375-8_2

1 Background and Theory

Machine Generated Keywords wavelet, wavelet transform, CQWT, quaternion wavelet, filter, continuous quaternion, Clifford, continuous, quadratic, rightside QFT, QFT, QWT, construct, rightside, equation

1.1 Background for QWT

Machine Generated Keywords wavelet, filter, Clifford, quadratic, wavelet transform, QWT, equation, solve, matrix, Clifford algebra, smooth, quaternion wavelet, optical flow, scaling, kind

1.1.1 QWT: Retrospective and New Applications

DOI: https://doi.org/10.1007/978-1-84996-108-0_13

Abstract-Summary
The quaternion wavelet transform (QWT) achieved much attention in recent years as a new image analysis tool.

It is an extension of the real wavelet transform and complex wavelet transform (CWT) by using the quaternion algebra and the 2D Hilbert transform of filter theory, where analytic signal representation is desirable to retrieve phase-magnitude description of intrinsically 2D geometric structures in a grayscale image.

A quaternionic phase congruency model is defined based on analytic signal representation so as to operate as an invariant feature detector for image registration.

To achieve better localization of edges and textures in image fusion tasks, we incorporate a directional filter bank (DFB) into the quaternion wavelet decomposition scheme to greatly enhance the direction selectivity and anisotropy of QWT.

The strong potential use of the QWT in color image recognition is materialized in a chromatic face recognition system by establishing invariant color features.

Acknowledgement
A machine generated summary based on the work of Xu, Yi; Yang, Xiaokang; Song, Li; Traversoni, Leonardo; Lu, Wei

2010 in Geometric algebra computing in engineering and computer science.

1.1.2 The Theory and Use of the Quaternion Wavelet Transform

DOI: https://doi.org/10.1007/s10851-005-3605-3

Abstract-Summary
This paper presents the theory and practicalities of the quaternion wavelet transform (QWT).

The major contribution of this work is that it generalizes the real and complex wavelet transforms and derives a quaternionic wavelet pyramid for multi-resolution analysis using the quaternionic phase concept.

For the estimation of motion through different resolution levels we use a similarity distance evaluated by means of the quaternionic phase concept and a confidence mask.

Acknowledgement
A machine generated summary based on the work of Bayro-Corrochano, Eduardo
2005 in Journal of Mathematical Imaging and Vision

1.1.3 Discrete Wavelets with Quaternion and Clifford Coefficients

DOI: https://doi.org/10.1007/s00006-018-0876-5

Abstract-Summary
We use paraunitary completion of the polyphase matrix to find corresponding quaternion and Clifford wavelet filters.

We then use the cascade algorithm on our filters to find quaternion and Clifford scaling and wavelet functions, which we illustrate using all possible projections onto two and three dimensions: to our knowledge, this is the first time that this has been done.

Acknowledgement
A machine generated summary based on the work of Fletcher, Peter
2018 in Advances in Applied Clifford Algebras

1.1.4 Quaternion-Valued Smooth Compactly Supported Orthogonal Wavelets with Symmetry

DOI: https://doi.org/10.1007/s00006-019-0945-4

Original Abstract We construct two novel quaternion-valued smooth compactly supported symmetric orthogonal wavelet (QSCSW) filters of length greater than existing ones. In order to obtain their filter coefficients, we propose an optimization-based method for solving a specific kind of multivariate quadratic equations. This method provides a new idea for solving multivariate quadratic equations and could be applied to construct much longer QSCSW filters.

Acknowledgement
A machine generated summary based on the work of Ma, Guangsheng; Peng, Lizhong; Zhao, Jiman
2019 in Advances in Applied Clifford Algebras

1.1.5 The Continuous Quaternion Algebra-Valued Wavelet Transform and the Associated Uncertainty Principle

DOI: https://doi.org/10.1007/s11868-021-00396-w

Abstract-Summary
This article aims to extend the wavelet transform to quaternion algebra using the kernel of the two-sided quaternion Fourier transform.

Introduction
Our contribution to these developments is that we introduce the two-dimensional continuous quaternion wavelet transform by means of the kernel of the two-sided QFT, and use the similitude group of the plane.

We thoroughly study this generalization of the continuous wavelet transform to quaternion algebra which we call the two-sided continuous quaternion wavelet transform (CQWT).

The novelty in the present work can be stated as follows: we use the quaternion split to prove a Plancherel identity for the two-sided QFT by means of the inner product and then follow the same process as in the Clifford case [1] as well as the CQWT case based on the kernel of the right-sided QFT [2], and construct our new transform and investigate its important properties such as linearity, scaling, rotation etc, we further show that these properties of the two-sided CQWT can be established whenever the quaternion wavelet satisfies a particular admissibility condition.

Uncertainty Principles for the Two-dimensional Continuous Quaternion Wavelet Transform
We prove the famous Heisenberg–Weyl's UP [3], and its logarithmic version for the CQWT.

That Heisenberg–Weyl's UP for the two-sided QFT states that a nonzero quaternion algebra-valued function and its QFT cannot both be sharply localized [4].

We extend the validity of this UP to the CQWT.

Conclusion
We developed the definition of the CQWT using the kernel of the two-sided QFT and the similitude group.

Using the derivative theorem, and Heisenberg–Weyl's UP related to the two-sided QFT, we derived Heisenberg–Weyl's UP associated with the CQWT.

Due to the spectral QFT representation of the CQWT and based on the logarithmic UP for the two-sided QFT, the associated forms with this UP have been proved in the CQWT domain.

Acknowledgement
A machine generated summary based on the work of El Haoui, Youssef
2021 in Journal of Pseudo-Differential Operators and Applications

1.2 Theoretical Developments for the QWT

Machine Generated Keywords CQWT, wavelet transform, wavelet, rightside QFT, rightside, continuous quaternion, continuous, quaternion wavelet, transform CQWT, QFT, commonly, linear canonical, smoothness, canonical, function

1.2.1 Linear Canonical Wavelet Transform in Quaternion Domains

DOI: https://doi.org/10.1007/s00006-021-01142-7

Abstract-Summary
For efficient analysis of such quaternionic signals, we introduce the notion of linear canonical wavelet transform in the quaternion domain by invoking the elegant convolution structure associated with the quaternion linear canonical transform.

We formulate three uncertainty principles: Heisenberg-type, logarithmic and local uncertainty inequalities associated with the linear canonical wavelet transform in the quaternion domain.

Introduction
With the development of the theory of linear integral transforms, the LCT is becoming increasingly popular and many ramifications have been witnessed in the open literature, including, linear canonical Gabor transforms [5], linear canonical wavelet transforms [6, 7], linear canonical Stockwell transforms [8] and many more [9, 10].

As of now, many integral transforms have been extended in the realm of quaternion algebra, including quaternion Fourier and wavelet transforms [11], quaternionic Ridgelet transform [12], quaternionic Stockwell transform [13], quaternionic curvelet transform [14], quaternionic shearlet transform [15] and many more [16, 17].

To address this limitation, our endeavour is to combine the merits of the well-known wavelet and linear canonical transforms in the realm of quaternion algebra by invoking the convolution theory associated with the quaternion linear canonical transform.

The main objectives of this article are as follows: To introduce the notion of linear canonical wavelet transform in the quaternion domain by invoking the convolution structure associated with the quaternion LCT.

To formulate certain uncertainty inequalities associated with the proposed linear canonical wavelet transform in the quaternion domain.

Preliminaries
Our aim is to present an overview of the quaternion algebra and the quaternion linear canonical transform which serves as a cornerstone for the subsequent developments.

The quaternion algebra provides an extension of the complex number system to an associative, non-commutative four-dimensional algebra.

There is a corresponding inversion formula. It is well-known that the quaternion multiplication is non-commutative, as such, there are three types of the quaternion linear canonical transforms (QLCT): the left-sided, right-sided and two-sided QLCT.

A Quaternion Analogue of Linear Canonical Wavelet Transform
We shall explicitly introduce the notion of linear canonical wavelet transform in the quaternion domain by formulating a new convolution structure associated with the two-dimensional linear canonical transform.

The corresponding generalized translation in the two sided QLCT domain is given. We now introduce a new convolution structure associated with the quaternion linear canonical transform and obtain the corresponding convolution theorem.

The new convolution structure shall be subsequently invoked to formulate the definition of the proposed linear canonical wavelet transform in the quaternion domain.

The next theorem guarantees the reconstruction of the input quaternion signal from the corresponding linear canonical wavelet transform in the quaternion domain.

Uncertainty Principles for Quaternion Linear Canonical Wavelet Transform
The classical Heisenberg uncertainty principle in harmonic analysis gives information about the spread of a signal and its Fourier transform by asserting that a signal cannot be sharply localized in both the time and frequency domain [18].

In analogy to the uncertainty principles governing the simultaneous localization of a function f and its Fourier transform, a different class of uncertainty principles comparing the localization of a function f with the localization of its Gabor or wavelet transform was studied by Wilczok [19].

Acknowledgement
A machine generated summary based on the work of Shah, Firdous A.; Teali, Aajaz A.; Tantary, Azhar Y.

2021 in Advances in Applied Clifford Algebras

1.2.2 The Quaternion Fourier and Wavelet Transforms on Spaces of Functions and Distributions

DOI: https://doi.org/10.1007/s40687-020-00209-4

Abstract-Summary
The continuous quaternion wavelet transform of periodic functions is also defined and its quaternion Fourier representation form is established.

Introduction
The inverse right-sided QFT of f is given. For various properties of the quaternion Fourier transformation, we may refer to [20, 2, 21, 22].

E. S. M. Hitzer has derived a directional uncertainty principle for quaternion valued functions subject to the quaternion Fourier transformation [23].

The quaternion Fourier transform is also used in image and signal processing [24, 25, 21].

[Section 2]
We have studied the right-sided QFT of test functions.

This shows the continuity of the right-sided QFT.

Continuous Quaternion Wavelet Transform of Periodic Functions
The continuous quaternion wavelet transform (CQWT) is a generalization of the classical continuous wavelet transform.

For various properties of the CQWT, we may refer to [26, 27, 28, 2, 29].

We have studied the continuous quaternion wavelet transform of periodic functions.

Acknowledgement
A machine generated summary based on the work of Lhamu, Drema; Singh, Sunil Kumar

2020 in Research in the Mathematical Sciences

1.2.3 Wavelet Transform of Dini Lipschitz Functions on the Quaternion Algebra

DOI: https://doi.org/10.1007/s00006-020-01112-5

Abstract-Summary
We generalize Titchmarsh's theorem for complex- or hypercomplex-valued functions.

Introduction
In [5], the authors have generalized the WFT to the windowed linear canonical transform (WLCT).

It is well-known that the quaternion Fourier transform (QFT), is a non trivial generalization of the real and complex Fourier transform (FT).

According to (right-sided) QFT, one can extend the classical wavelet transform (WT) to quaternion algebra, while having the same properties as in the classical case.

It should also be mentioned that this wavelet transform uses the kernel of the (right-sided) QFT which in general does not commute with quaternions.

A version of Titchmarsh's theorem and Dini Lipschitz Functions for the Quaternion Fourier transform (QFT) and the Quaternion Linear Canonical transform (QLCT) is given, as it appears in [30, 31].

The purpose of this paper is to study the Wavelet Transform of Dini Lipschitz Functions on the Quaternion Algebra.

Acknowledgement
A machine generated summary based on the work of Bouhlal, A.; Safouane, N.; Achak, A.; Daher, R.
2021 in Advances in Applied Clifford Algebras

1.2.4 Equivalence Between K-Functionals and Modulus of Smoothness on the Quaternion Algebra

DOI: https://doi.org/10.1007/s11565-022-00387-9

Introduction
Using the (right-sided) QFT, one can extend the classical wavelet transform (WT) to quaternion algebra, while having the same properties as in the classical case.

It should also be mentioned that this wavelet transform uses the kernel of the (right-sided) QFT which in general does not commute with quaternions.

It is well-known that studying the relation which exists between the smoothness properties of a function and the best approximations of this function in weight functional spaces is more convenient than usual with various generalized moduli of smoothness (see [32, 33]).

The study of the relation which exists between the modulus of smoothness and K-functionals is known as one of the major problems in the theory of the approximation of functions.

Acknowledgement
A machine generated summary based on the work of Bouhlal, A.; Safouane, N.; Belkhadir, A.; Daher, R.
2022 in Annali dell' Università di Ferrara

2 Image Processing

Machine Generated Keywords wavelet, QWT, wavelet transform, phase, denoise, image denoise, quaternion wavelet, complex wavelet, patch, filter, coefficient, denoising, Gaussian, multiscale analysis, multiresolution

2.1 Image Processing, Colorization and Sharpness

Machine Generated Keywords patch, multiresolution, phase, wavelet, distribution, quaternion wavelet, filter, level, wavelet transform, metric, object, color, use quaternion, similar, Gaussian

2.1.1 Multi-Resolution Image Analysis Using the Quaternion Wavelet Transform

DOI: https://doi.org/10.1007/s11075-004-3619-8

Abstract-Summary
This paper presents the theory and practicalities of the quaternion wavelet transform.

The contribution of this work is to generalize the real and complex wavelet transforms and to derive for the first time a quaternionic wavelet pyramid for multi-resolution analysis using the quaternion phase concept.

The three quaternion phase components of the detail wavelet filters together with a confidence mask are used for the computation of a denser image velocity field which is updated through various levels of a multi-resolution pyramid.

Acknowledgement
A machine generated summary based on the work of Bayro-Corrochano, Eduardo 2005 in Numerical Algorithms

2.1.2 Quaternion Atomic Function for Image Processing

DOI: https://doi.org/10.1007/978-0-85729-811-9_8

Abstract-Summary
We introduce a new kernel for image processing called the atomic function.

We discuss the role of the quaternion atomic function with respect to monogenic signals.

Making use of the generalized Radon transform and images processed with the quaternion wavelet atomic function transform, we detect shape contours in color images.

We believe that the atomic function is a promising kernel for image processing and scene analysis.

Introduction
This work introduces the atomic function for grey scale and color image processing using the quaternion algebra framework.

This chapter shows how quaternion atomic filters can be used as smoothing and differentiator filters to carry out differential geometry on the visual manifold.

The quaternion atomic filters appear to be useful for differential geometry.

This work presents the theory and some applications of the quaternion atomic function Qup in image processing as a novel quaternionic wavelet.

We developed the Qup(x) using the quaternion algebra H. The quaternion atomic function permits us to extract the phase information from the image [34, 25, 35].

Using the generalized Radon transform and images processed with the quaternion wavelet atomic function, we look for contours in color images, showing that the atomic function is a promising kernel for image processing and scene analysis.

Atomic Functions

In the AF class, the function up(x) is the simplest and, at the same time, most useful primitive function to generate other kinds of atomic functions [36].

The function up(x) is infinitely differentiable, $up(0) = 1$, $up(-x) = up(x)$.

The atomic function up(x) is generated by infinite convolutions of rectangular impulses.

Atomic windows were compared with classic ones [37, 36] by means of a system of parameters such as the equivalent noise bandwidth, the 50% overlapping region correlation, the parasitic modulation amplitude, the maximum conversion losses (in decibels), the maximum side lobe level (in decibels), the asymptotic decay rate of the side lobes (in decibels per octave), the window width at the six-decibel level, and the coherent gain.

This problem reduces to the synthesis of infinitely differentiable finite functions with small-diameter carriers that are used for constructing the weighting windows [37, 38, 39].

We abbreviate this family of derivative functions as dup(x).

Quaternion Algebra

The quaternion algebra H was invented by W.R. Hamilton in 1843 [40].

It is an associative, non-commutative, four-dimensional algebra with orthogonal imaginary numbers i, j, and k. The conjugate of a quaternion is given. For the quaternion q, we can compute its partial angles and its partial moduli and its projections on its imaginary axes. The concept of a quaternionic Hermitian function is very useful for the computation of any kind of inverse quaternionic transforms using the quaternionic analytic signal.

If $e^{i\varphi} e^{k\psi} e^{j\theta} = -q$ and $\varphi < 0$, then $\varphi \rightarrow \varphi + \pi$.

[Section 4]

The up(x) function is easily extendible to two dimensions.

We show the performance of Qup to detect edge changes using the phase concept.

We can see that the phases yield very useful information about shape contour changes.

Monogenic Signal and the Atomic Function

Using the same embedding, the monogenic signal can be defined in the frequency domain where the inverse Fourier transform of $F_M(u)$ is given. $f_R(x)$ stands for the Riesz transform [41] obtained by taking the inverse transform of $F_R(u)$. $\star$ stands for the convolution operation.

Using the fundamental solution of the 3D Laplace equation restricted to the open half-space $z > 0$ with boundary condition, the solution is defined, and h_P stands for the 2D Poisson kernel.

Setting in $f_M(x,y,z)$ the variable z equal to zero, we obtain the so-called monogenic signal.

The monogenic functions are the solutions of generalized Cauchy–Riemann equations or Laplace-type equations.

Quaternion Wavelet Atomic Function Transform

Thanks to the development of the quaternion Fourier transform (QFT) [35], the generalization of the real and complex wavelet transform to the Quaternion Wavelet Transform (QWT) was straightforward.

The QWT is a natural extension of the real and complex wavelet transform, taking into account the axioms of quaternion algebra, the quaternionic analytic signal [35], and its separability property.

Multi-resolution analysis can also be straightforwardly extended to the quaternionic case; we can therefore improve the power of the phase concept, which is not possible in real wavelets and, in the case of complex wavelets, is limited to only one phase.

We use this phase information in the quaternionic wavelet multi-resolution analysis.

The quaternionic atomic function wavelet analysis from level $\alpha - 1$ to level α corresponds to the transformation of one quaternionic approximation to a new quaternionic approximation and three quaternionic differences. Note that we do not use the idea of a mirror tree [34].

Radon Transform of Functionals

The Radon transform (RT), introduced by J. Radon in 1917, describes a function in terms of its (integral) projections [42].

The RT can be seen as the mapping from the function onto the projections of the Radon transform.

The original formulation of the RT represents the projections of the function I along the lines $c_l(d,\varphi)$.

The inverse of the RT corresponds to the back-reconstruction of the function from the projections.

The RT builds in the Radon parameter space a function P(p) having peaks for those parameter vectors p for which the corresponding shape c(p) is present in the image.

[Section 8]

In certain levels of the multiresolution pyramid, one can compute the RT for extracting shapes.

This experiment clearly shows how the phase information can be used to localize and extract more information independently of the contrast of the image.

The Qup kernel was used as the mother wavelet in the multi-resolution pyramid.

The Qup mother wavelet kernel was steered: elongation $s_x = 0.3$ and $s_y = 0.25$ and angles {0°, 22.5°, 45°, 77.25°, 90°} through the multi-resolution pyramid.

The image was then filtered by steering a quaternion wavelet atomic function, and at a certain level of the multi-resolution pyramid we applied the RD.

One can improve performance of the multi-resolution approach by applying the RT from coarse to fine levels to identify possible shape contours in the upper level.

Conclusion
This paper introduces the theory and some applications of the quaternion atomic function wavelet in image processing.

Making use of the generalized Radon transform and images processed with the quaternion wavelet atomic function transform, we also detect shape contours in color images.

We hope that this work will encourage computer scientists and practitioners to use quaternion wavelet atomic functions to tackle various problems in image processing and scene analysis.

Exercises
Derive a sort of quaternion Harris detector for color image processing using the quaternion atomic filter which only detects corners of a selected color.

Derive an algorithm for the RT using color images and quaternion atomic filters to detect lines, circles and ellipsoids of one particular color.

Apply a conformal transformation to map the color image plane to the sphere, then using a quaternion atomic filter derive a line and circle detector on the sphere which detects with respect to only a particular color.

Acknowledgement
A machine generated summary based on the work of Bayro-Corrochano, Eduardo; Moya-Sánchez, Eduardo Ulises

2011 in Lecture Notes in Computer Science

2.1.3 Colorization Using Quaternion Algebra with Automatic Scribble Generation

DOI: https://doi.org/10.1007/978-3-642-27355-1_12

Abstract-Summary
In current colorization techniques, major user intervention is required in the form of tedious, time-consuming scribble drawing.

We focus on automatic scribble generation and a structure-preservation mechanism, which are still open issues of colorization.

We generate scribbles automatically along points where the spatial distribution entropy achieves locally extreme values.

Given the color scribbles, we compute quaternion wavelet phases to conduct colorization along equal-phase lines.

The experimental results demonstrate that the proposed colorization method can achieve natural color transitions between different objects with automatically generated scribbles.

Acknowledgement
A machine generated summary based on the work of Ding, Xiaowei; Xu, Yi; Deng, Lei; Yang, Xiaokang

2012 in Lecture Notes in Computer Science

2.1.4 The Image Sharpness Metric via Gaussian Mixture Modeling of the Quaternion Wavelet Transform Phase Coefficients with Applications

DOI: https://doi.org/10.1007/978-3-319-26396-0_13

Abstract-Summary
Derived from the parameters in GMM, the metric is proposed to find the relationship between the image blur degree and the distribution histograms of high frequency coefficients.

Experiments are conducted on natural images and the reasonable results indicate that the proposed metric can exhibit better performance than three common global sharpness measurements and satisfy the visual perception in the local smooth patch detection.

Extended:

The comparative study on different mixture models (e.g. Laplacian mixed with Gaussian model) and the further error analysis will be done in future work.

Introduction
Quaternion wavelet transform (QWT), as a novel image analysis tool, has some superior properties compared to discrete wavelets, such as nearly shift-invariant wavelet coefficients and the ability of texture presentation which provides richer image texture information because of the phases.

Gai and Liu use a hidden Markov model (HMM) [43] and a Copula model [44] to exploit the relationship between magnitude and phases of the QWT with application to texture classification, respectively.

We explore the potential of the Gaussian mixture model (GMM) inspired by the bimodal characteristics of phase coefficients.

The basic idea underlying our method is that we derive the metric by means of multiple Gaussian parameters to depict the relationship between phase change and blur degree by means of analyzing the histogram of the QWT coefficients of natural images.

Quaternion Wavelet Transform
For further discussions, we briefly review some basic ideas on quaternions and the construction of the QWT.

According to the definition of the quaternionic analytic signal, the QWT, i.e. the analytic 2D wavelets can be constructed.

Each sub-band of the QWT can be seen as an analytic signal associated with a narrow band part of the image.

More details about implementation of the QWT used here are referenced in the work [45].

The Proposed Metric via Gaussian Mixture Model
The expectation maximization (EM) algorithm is expected to estimate the probabilistic model parameters with hidden variants.

By virtue of multiple mixture models, it is expected to estimate the complex coefficients distribution.

Imaging systems often suffer from blur owing to defocus effects.

Experimental Results

We analyze the effect of the number of GMMs and apply the proposed metric in two areas: Global image sharpness detection and local image smooth patches detection.

Smooth patch detection plays a key role to noise level estimation, segmentation, object extraction, etc. Actually, smooth patches are similar to the seriously blurred ones with less textures/features/edges.

Sometimes a local patch does not have enough pixels for estimating the GMM parameters, so we only use the coefficient variance with values greater than 0 to substitute the GMM results, meanwhile the displayed black squares have relatively smaller metric values that means relatively smooth patches.

Conclusions

We propose one image sharpness metric based on the coefficient statistics of the QWT phases.

The metric is derived from the GMM parameters of the phase coefficient distribution to describe the blur degree.

The novelty of this paper comes from introducing the GMM into modeling the QWT phase coefficients.

The magnitude-phase representation form is computed before estimating GMM parameters for QWT phase coefficients.

Acknowledgement

A machine generated summary based on the work of Liu, Yipeng; Du, Weiwei 2015 in Studies in Computational Intelligence

2.2 *Image Denoising and Saliency*

Machine Generated Keywords wavelet, denoise, image denoise, QWT, wavelet transform, denoising, multiscale analysis, descriptor, saliency, image denoising, complex wavelet, remove, noise, coefficient, dualtree

2.2.1 Image Denoising via Modified Multiple-Step Local Wiener Filter and Quaternion Wavelet Transform

DOI: https://doi.org/10.1007/978-3-662-49831-6_76

Abstract-Summary

This paper proposes an image denoising algorithm via the modified multiple-step local Wiener filter in the quaternion wavelet transform domain.

The multiple-step local Wiener filter has shown good performance on removing Gaussian white noise.

Experimental results verify that the proposed method improves the denoising performance significantly and is very efficient in computation time.

Extended:

The multiple-step local Wiener filter [46] is modified to shrink the noisy coefficients.

Introduction

Transform domain methods by using wavelets do well in removing the noise of images.

Although the wavelet transform has been widely used in the design of image denoising, wavelets are only very efficient in dealing with point singularities.

Many new wavelets are proposed and used in image processing.

The quaternion wavelet transform (QWT) proposed by Corrochano [47, 16] is an effective multiscale analysis tool in image processing.

In [48], Shan and others proposed a new image denoising algorithm via bivariate shrinkage based on the QWT (QWT-Bishrink).

We note that the diffusion scheme has been introduced into the wavelet domain for image denoising unlike the traditional wavelet shrinkage strategy.

In this paper, we also propose an image denoising algorithm.

We also compare the proposed method with other related image denoising methods.

The Proposed Method

The QWT can be implemented with separable filter banks and is carried out by using a dual-tree algorithm with linear computational complexity [46].

Compared to the wavelet transform, the QWT is shift invariant and has good directional selectivity.

Compared to the complex wavelet transform, the QWT has abundant phases capturing the geometric structure of natural images.

We summarize the algorithm as follows: (1) Perform the QWT on the original noisy image. (2) Modify the QWT coefficients by using a modified multiple-step local Wiener filter. (3) The noise-free image is finally estimated by computing the inverse QWT.

Experimental Results

In terms of [49], the SURE-LET method outperforms some state-of-the-art wavelet-based denoising methods, such as BayesShrink, ProbShrink, BiShrink, etc. For this, in this paper, we focus on the comparison between the proposed method and the SURE-LET method.

For the proposed method, a QWT with five-level decomposition is used.

Among all cases, the proposed method obtains higher PSNRs.

Compared to the SURE-LET approach, the proposed method can save more time.

The proposed method is more efficient and effective when dealing with noisy images.

Conclusion

It is based on an integration of the multiple-step local Wiener filter and the QWT.

This method can be considered as an extension and supplementary to a diffusion-based Wiener filter.

The effectiveness of the multiple-step local Wiener filter is further realized.

Acknowledgement

A machine generated summary based on the work of Zhang, Xiaobo 2016 in Lecture Notes in Electrical Engineering

2.2.2 Image Denoising Using Normal Inverse Gaussian Model in Quaternion Wavelet Domain

DOI: https://doi.org/10.1007/s11042-013-1812-2

Abstract-Summary

The proposed algorithm is based on a design of a Maximum Posteriori Estimator (MAP) combined with a Quaternion Wavelet Transform (QWT) that utilizes the Normal Inverse Gaussian (NIG) Probability Density Function (PDF).

A NIG PDF which is specified by four real-value parameters is capable of modeling the heavy-tailed QWT coefficients, and describing the intra-scale dependency between the QWT coefficients.

The NIG PDF is applied as a prior probability distribution, to model the coefficients by utilizing the Bayesian estimation technique.

A simple and fast method is given to estimate the parameters of the NIG PDF from neighboring QWT coefficients.

Introduction

A Gaussian scale mixture model based on complex wavelet transform coefficients for image denoising was proposed [50].

M.N. Do [51] proposed a new image denoising methodology based on the Contourlet Transform.

Cunha [52] proposed a new image denoising algorithm based on a non-subsampled Contourlet Transform which could be considered to be an improved Contourlet Transform.

The drawbacks of denoising methods based on the complex wavelet transform and the Contourlet Transform, are a lack of phases that can capture the geometric structure of the natural images.

The QWT [16, 47], based on the theory of the 2-D Hilbert Transform, has recently emerged as a new multi-scale analysis tool for signal and image processing.

Statistical approaches focused on the development of statistical models for natural images and their transform coefficients, have recently emerged as a new tool for image denoising.

Preliminaries

The three angles (ϕ,θ,ψ), called phase angles, are calculated. A Quaternion Wavelet Transform (QWT) has one magnitude and three phase angles.

A QWT is considered to be a local Quaternion Fourier transform (QFT).

The quaternionic analytic signal with a 2D function is defined. H_{i1} and H_{i2} are partial Hilbert Transforms respectively, H_i is a total Hilbert Transform, and f(x,y) is a real-valued, 2-D signal.

The mother wavelet is a 2D quaternionic analytic filter that can generate coefficients that are analytic.

The analytic extension in the quaternion domain is defined. From a practical perspective, if the mother wavelet is separable, then the 2D Hilbert Transform is equivalent to a 1D Hilbert Transform of rows and columns.

The 2D QWT is written in terms of separable products. The decomposition is dependent on the position of the image with respect to the x and y axes.

Normal Inverse Gaussian Density Model

As the QWT coefficients are generally non-Gaussian and have heavy-tailed data, the NIG PDF is suitable for modeling the QWT coefficients of the log transformed image.

NIG PDF, Laplacian PDF and Generalized Gaussian PDF (GG PDF) are fitted to this density.

The NIG PDF can fit the QWT coefficients better than the other PDFs.

The NIG PDF was selected as the prior model of the QWT coefficients in the proposed algorithm attributes for two reasons: first, the selection of the distributed parameters of NIG PDF is flexible; and second, the QWT coefficients corresponding to image edges demonstrate sharp peak values and are heavy tailed for non-Gaussian characteristics.

Proposed Method

In the proposed method, the denoising of an image is carried out in the Quaternion Wavelet domain by means of a Bayesian MAP estimator.

The denoising problem in the QWT domain can be formulated. Q(i, j) is a noisy QWT coefficient, and X(i, j) and N(i, j) denote true QWT coefficients and noise, respectively.

The pixel position of the natural images is denoted by (i, j).

We apply the inverse QWT to the denoised quaternion wavelets and obtain the denoised image.

Experimental Results

The third experiment was conducted to illustrate the edge preservation abilities of the different methods.

The Edge Preservation Index B, which is utilized for evaluating the ability of a denoising method to preserve the edge, and the Structural Similarity (SSIM), which is used for assessing the visual quality of the denoised image, were both applied in this paper for evaluating the edge preservation abilities.

It is clear to see that the QNIG method can provide better edge presentation of the edges in denoised images.

The purpose of the fourth experiment was to illustrate the effectiveness of the proposed method, as compared with other denoising methods, by using different noise levels.

Although the elapsed times of WF, BS and BM3D were less than that of the QNIG, the denoising effects for each method were below those for the QNIG.

Conclusion

We have proposed a QWT based image denoising method.

The quintessential step of the proposed method was determining the most appropriate prior model, which was then utilized in describing the marginal distribution of the QWT coefficients and in applying the parameter estimation method.

The parameters of the model were estimated using the Intra-Scale Neighboring QWT coefficients.

In denoising, the proposed combination of the QWT method with NIG density was performed on the Peppers, Boat, Camera, Lena and Barbara test images, respectively.

Acknowledgement

A machine generated summary based on the work of Gai, Shan; Luo, Limin 2013 in Multimedia Tools and Applications

2.2.3 Image Denoising Using Fractional Quaternion Wavelet Transform

DOI: https://doi.org/10.1007/978-981-10-7898-9_25

Abstract-Summary

This work presents an image denoising algorithm using the fractional quaternion wavelet transform (FrQWT).

Images corrupted with additive Gaussian noise are considered and a FrQWT is performed via hard and semi-soft thresholds.

The thresholding on the wavelet coefficients reveals the capabilities of the wavelet transform in the restoration of an image degraded by noise.

Extended:

We will include control parameter sensitivity of the hard threshold so that we can have better qualitative and quantitative improvement in recovered images.

Introduction

Wavelet analysis is coming out as one of the most important tools in signal analysis, pattern recognition, image processing and other fields [53].

The wavelet transform breaks down an image into different frequency and space sub-images, and the coefficients of these sub-images are then processed.

Progressively, the dual-tree complex wavelet transform or in short ‘CDWT’ was proposed by Kingsbury [54, 55].

It was basically a computational structure of the complex wavelet transform (CDWT) which has been used frequently in various image processing applications [56].

The quaternion wavelet transform (QWT) can be considered as a novel multi-scale analysis tool.

The fractional quaternion wavelet transform (FrQWT) is proposed in a similar manner as the QWT [57].

The FrQWT is used to transform the noisy image in the frequency domain as a quadrature of different sub-bands.

Preliminaries

The wavelet transform [53] can be defined as the decomposition of a signal with a family of real orthonormal bases.

The smaller wavelet coefficients usually represent the noise in the image as opposed to the coefficients with a larger magnitude value which contain more signal information than noise.

By removing the noisy (smaller) coefficients and taking the inverse wavelet transform may obtain a reconstruction that has a reduced amount of noise.

The real and imaginary components from the wavelet coefficients are used to calculate the magnitude information and phase angle.

FrQWT: Definition and Implementation

The FrQWT is defined in a similar fashion as the QWT [57].

We make use of the fractional Hilbert operator in place of the classical Hilbert operator to generalize the QWT in the FrQWT.

A 2-D implementation of a dual-tree filter bank is used separately for computing the FrQWT coefficients.

Image Denoising Using FrQWT

We are using thresholding on the FrQWT coefficients to remove the noise.

We are also using phase regularization on thresholded coefficients to further denoise the image.

The process of thresholding can be considered as a simple nonlinear filtering that operates on one wavelet coefficient at a time [58].

The choice of a large threshold results in the loss of texture details from the image.

The thresholding use is to be done only on the magnitude of the FrQWT coefficients.

This particular choice of regularizer enforces spatial smoothness in the transformed coefficients, which is required after the thresholding-based removal of the noise from the images.

The transformed signal can be represented in polar form with the following expression: Here, the regularization of the third-phase angle together with a thresholding on the magnitude of the fractional quaternion wavelet coefficients results in a perfect technique for image denoising.

Experiment Results and Analysis

The first column describes the image, while the second column gives the PSNR value between original (ground truth) and their respective noisy image.

The aim of denoising is to increase this PSNR between the restored and original image.

In case of the cameraman image, the performance of the FrQWT was found better than other variants of wavelets.

In the rest of the images, the performance in the increasing order can be seen as DWT, CDWT, QWT and FrQWT.

The other observation is that in most of the images, semi-soft thresholding does better work when compared to hard thresholding.

In the second set of images with noise of variance 0.01, the performance of the FrQWT is found better than DWT, CDWT and QWT in case of all the images.

The semi-soft scheme of thresholding performed marginally better than hard thresholding.

Conclusions

Further, it has been found that the choice of hard or semi-soft thresholdings depends on the image.

In some of the images, semi-soft threshold performs marginally better than hard threshold.

In some cases, it was found that the choice of a hard threshold is better.

We will include control parameter sensitivity of the hard threshold so that we can have better qualitative and quantitative improvement in recovered images.

Acknowledgement

A machine generated summary based on the work of Nandal, Savita; Kumar, Sanjeev 2018 in Advances in Intelligent Systems and Computing

2.2.4 Information-Based Scale Saliency Methods with Wavelet Sub-band Energy Density Descriptors

DOI: https://doi.org/10.1007/978-3-642-36543-0_38

Abstract-Summary

Pixel-based scale saliency (PSS) work is based on information estimation of data content and structure in multiscale analysis; its theoretical aspects as well as practical implementation are discussed by Kadir et al [11].

The Scale Saliency framework [10] does not work only for pixels but also other basis-projected descriptors as well.

Our contribution is introducing a mathematical model of utilizing wavelet-based descriptors in a correspondent Wavelet-based Scale Saliency (WSS).

It treats wavelet sub-band energy density of the popular discrete wavelet transform (DWT) and the dual-tree complex wavelet transform (DTCWT) as basis descriptors instead of pixel-value descriptors for saliency map estimation.

Acknowledgement
A machine generated summary based on the work of Le Ngo, Anh Cat; Ang, Li-Minn; Qiu, Guoping; Seng, Kah Phooi
2013 in Lecture Notes in Computer Science

3 Watermarking and Forensic Features

Machine Generated Keywords scheme, watermarking, QWT, watermark, attack, algorithm, DCT, watermarke, watermarking scheme, chaotic, transform DCT, robust, statistical, image watermarking, digital

3.1 A Novel Color Image Watermarking Algorithm Based on QWT and DCT

DOI: https://doi.org/10.1007/978-981-10-7299-4_35

Abstract-Summary
A novel color image watermarking algorithm is proposed based on the quaternion wavelet transform (QWT) and the discrete cosine transform (DCT) for copyright protection.

The luminance channel Y of the host color image in YCbCr space is decomposed by the QWT to obtain four approximation subimages.

Introduction
Image watermarking algorithms can be classified into spatial domain algorithms and transform domain algorithms.

For the spatial domain algorithms, the watermark is usually embedded into the host image by modifying its pixels directly [59, 60].

For transform domain watermarking algorithms, the host image is firstly decomposed by some transforms, and then the watermark is embedded by modifying the coefficients after the transformation.

The popular transform domain watermarking algorithms include Discrete Cosine Transform (DCT) [61], Discrete Fourier Transform (DFT) [62], Discrete Wavelet Transform (DWT) [63], Discrete Shearlet Transform (DST) [64] and so on.

A watermarking scheme using subsampling based on the DCT was proposed by Chu and others [65].

Many subsampling-based watermarking algorithms have been developed.

A novel image watermarking scheme is proposed based on the QWT and the DCT.

The processes of watermark embedding and extracting are implemented on the luminance component of a host color image in the YCbCr space.

Quaternion Wavelet Transform
The QWT overcomes the common drawbacks of wavelet transforms by its shift-invariant feature.

Besides of the shift-invariant magnitude, the QWT has three phases.

The first two phases represent local image shifts, and the third one denotes the texture information [66].

Watermarking Scheme Based on DCT and QWT
The novel watermarking embedding and extracting schemes are detailed based on the DCT and the QWT.

The steps of embedding a watermark into the original host color image are described.

The detailed steps of watermark extracting are described.

Experimental Results
Imperceptibility means that the human visual quality of the original host image should not be affected much even after watermark embedding [61].

If the NC value is near to 1, it means that the extracted watermark is strongly correlated to the original watermark.

If the similarity between the original watermark and the extracted watermark is higher, the NC value is larger, which indicates that the algorithm is more robust.

This can be proved by extracting the watermark image and calculating the NC between the original watermark and the extracted watermark after image attacks.

Conclusion
A novel color image watermarking algorithm has been proposed based on the QWT and the DCT.

The experimental results have shown that the proposed method attains more excellent imperceptibility and has stronger robustness against some attacks than the watermarking schemes in [67, 68, 69].

We will extend the proposed watermarking scheme to video watermarking.

Acknowledgement
A machine generated summary based on the work of Han, Shaocheng; Yang, Jinfeng; Wang, Rui; Jia, Guimin

2017 in Communications in Computer and Information Science

3.1.1 A Novel Robust Image Watermarking in Quaternion Wavelet Domain Based on Superpixel Segmentation

DOI: https://doi.org/10.1007/s11045-020-00718-z

Abstract-Summary

Based on superpixel image segmentation and the quaternion wavelet transform, we propose a digital image watermarking approach which is highly robust against common image processing operations and local geometric transformations.

For each segment region, image feature points are extracted using the SIFER detector, and the affine invariant local regions are constructed adaptively.

The digital watermark is embedded into the local regions by modulating the invariant modulus coefficients.

Experimental results are provided to illustrate the efficiency of the proposed image watermarking, especially for noise attacks and local desynchronization attacks.

Extended:

Future work will consider using more CMG filters at 45 degrees to get planar rotational invariance as explained in the reference (Mainali 70).

Introduction

As for robust image watermarking methods, the feature-based watermarking methods have superior performance than other approaches in terms of overall robustness.

Yang and others 71 proposed a robust image watermarking scheme against local geometric distortions based on a SURF detector and invariant radial harmonic Fourier moments.

Through in-depth research and analysis, we observed that there are some important drawbacks in current feature-based image schemes: First, the feature points focus too much on the high contrast region, and the feature points are distributed unevenly.

To solve the contradiction between robustness and imperceptibility of the watermark, we use the quaternion wavelet transform (QWT) (Li and others 44; Fletcher and Sangwine 72) in this study.

The QWT can overcome two defects of discrete wavelet: (1) After the image shifts slightly, the characteristics of the image in the smooth area and the edge area will have a lot of change, such as that the edge area becomes blurred. (2) It can provide phase that represents the local information of the image.

Entropy Rate Superpixel Segmentation

Through superpixel segmentation, we can subdivide the image into many uniform regions.

In image classification, entropy is a measure of the image information.

The concept of entropy can be used to measure the complexity of texture in an image.

The larger the entropy, the more information, and the more change of the corresponding image part, which contains the texture and edge parts.

Superpixels segment the original image to get the superpixel blocks, and calculate the entropy of each superpixel.

The image blocks are divided into smooth and non-smooth regions in light of the average entropy.

If the entropy of the image block is greater than the average entropy then it is a non-smooth region.

Or the entropy of the image block is smaller than the average entropy then it is a smooth region.

Feature Points Extraction and Image Regions Construction
A novel method that extracts feature points based on a CM-Gaussian filter has been proposed by Mainali and others 73.

In light of the uncertainty spread theory of Heisenberg, the SIFER detector based on a CM-Gaussian filter has the smallest spread in both scale and space, since the SIFER detector needs band-pass filters to respond on a wide range of scales.

We use the SIFER detector based on a CM-Gaussian filter to detect the feature points.

In the SIFER algorithm, we can filter out the image features of different granularity, and strongly resist image artifacts at the same time.

Quaternion Wavelet Transform
The Quaternion wavelet transform (QWT) is a developing frequency domain analysis tool that provides the phase and magnitude information that enables some new applications in the field of image analysis and processing.

The real discrete wavelet transform is not shift-invariant, and has only one phase information.

The QWT is nearly shift-invariant and provides one magnitude and three phase information in different scales.

The QWT is composed of four real discrete wavelet transforms, the first real discrete wavelet transform corresponds to the real part of the quaternion wavelet, the other three real discrete wavelet transforms are obtained using the partial Hilbert transform and the total Hilbert transform.

We select the magnitude and phase of the QWT to describe the image information.

Watermark Embedding Scheme
Mapping a Local Elliptical Region to a Local Circular Region in the mapping, the circular area cannot be larger than the elliptical area.

The watermarked local circular region can be obtained by a zero-removing operation.

The watermarked local circular region is inverse-mapped into the watermarked local elliptical region.

Watermark Detection Scheme
If there are at least two local elliptical feature regions that are detected as "success", the image is claimed to be watermarked.

When at least two local elliptical feature regions are claimed watermarked, the final detection is claimed a "success".

Simulation Results

When the size of the quantization step is larger, the robustness performance of the proposed watermarking algorithm is increased, but imperceptibility performance will be reduced at the same time.

To quantitatively evaluate the invisibility performance of the proposed watermarking algorithm, we also calculate the PSNR, which is an objective criterion and is always used to evaluate image quality.

We use different image resolutions ranging from 512 × 512 × 8 bit to 1024 × 1024 × 8 bit, and compute the corresponding PSNR, the corresponding CPU time of watermark embedding/extracting, and the corresponding watermark detection rate under several attacks by utilizing the test image Lena.

When the image resolution is larger, the imperceptibility performance of the proposed watermarking algorithm is reduced, the robustness performance is changed slightly, but the average watermark embedding/extracting time will increase at the same time.

Conclusion

Desynchronization attacks, especially local ones that cause displacement between embedding and detection, are usually difficult for the watermark to survive.

We have provided a novel scheme using entropy rate superpixel segmentation and local QWT.

To resist the local desynchronization attacks and embed a watermark more easily, we build the local elliptical feature regions and map a local elliptical region to a circular region.

Experimental results are provided to illustrate the efficiency of the proposed image watermarking, especially for noise attacks and local desynchronization attacks.

Acknowledgement

A machine generated summary based on the work of Niu, Pan-pan; Wang, Li; Shen, Xin; Zhang, Si-yu; Wang, Xiang-yang

2020 in Multidimensional Systems and Signal Processing

3.1.2 Optimal Image Watermarking Scheme Based on Chaotic Map and Quaternion Wavelet Transform

DOI: https://doi.org/10.1007/s11071-014-1634-4

Abstract-Summary

An intelligent watermarking scheme optimized by adaptive differential evolution (ADE) is proposed based on a chaotic map and the quaternion wavelet transform (QWT).

The ADE algorithm is explored to optimize the watermarking parameters (i.e., scaling factors) automatically.

Comparison results also indicate the superiority of the proposed algorithm over existing watermarking algorithms.

Extended:

An intelligent watermarking scheme for image copyright protection is proposed based on the chaotic map, SVD and QWT.

The future direction of this work can be extended to address non-differentiable signals defined on Cantor sets using local fractional discrete or continuous wavelets.

Introduction

To achieve the balance of these constraints, artificial intelligence-based approaches such as particle swarm optimization (PSO), differential evolution (DE), and genetic algorithms have been widely applied in recent years with demonstrated remarkable performance for watermarking application [74–75].

The integration of SVD with the transform approach is able to achieve very remarkable watermarking performance [76, 75, 77].

A digital image watermarking scheme based on the combination of QWT and SVD has been investigated to insert watermarks based on a chaotic map.

The contribution of this paper includes the following: (1) Robustness against attacks especially geometrical distortions by the shift invariance of the QWT and merits of SVD; (2) High level of security using PWLCM encryption of a watermark before embedding; (3) Optimal balance of watermarking requirement is obtained with ADE training and an optimization and uniquely designed objective function.

Proposed Method

For the embedding and extraction rule, the spread spectrum method is explored due to its simplicity, effectiveness and promising performance for watermarking application.

Using chaotic scrambling, the watermark image is transformed into a disordered and meaningless sequences.

Without the scrambling algorithm and corresponding key, an attacker is not able to obtain the copyright watermark even if the watermark is extracted from the watermarked image.

A small shift does not change the coherence representation of the wavelet transform, and hence, it is very desirable for the watermarking application.

Most of the existing watermarking schemes are implemented by manually tuned constant values, but this watermarking scheme is actually host image dependent.

An algorithm is needed to obtain the optimum performance automatically to have a good trade-off between watermarking fidelity and robustness.

The objective function aims to optimize the watermarking performance of robustness, imperceptibility and capacity.

Experimental Results

The watermarking scheme obtains remarkable results under numerous common signal processing attacks especially for geometrical distortions.

The proposed algorithm resists geometrical distortions such as rotation, scaling and translation.

Geometric processing attacks, including image rotation, scaling, translation and cropping, are mainly modification of the pixel positions or orders without change of the image brightness and contrast.

Geometric distortion tests are very important for robustness performance evaluation of a watermarking scheme.

To further evaluate the robustness of the proposed algorithm, common geometric distortions are carried out to demonstrate the attractive property of the QWT and SVD especially for shift invariance.

It can be seen that the proposed watermarking scheme is very robust against the image translation attack.

Good statistics of the QWT, SVD and ADE have a positive effect on the analysis compared with the traditional least significant bit watermarking algorithm utilized in [78].

Conclusions

An intelligent watermarking scheme for image copyright protection is proposed based on the chaotic map, SVD and QWT.

The proposed method inserts watermarks into the original host data by modifying the singular values of the QWT amplitude coefficients.

Shift invariance of the QWT and desirable features of the SVD make the watermarking scheme highly robust against geometric operations.

The proposed algorithm shows superiority over related algorithms too.

Acknowledgement

A machine generated summary based on the work of Lei, Baiying; Ni, Dong; Chen, Siping; Wang, Tianfu; Zhou, Feng

2014 in Nonlinear Dynamics

3.1.3 A Robust Color Image Watermarking Algorithm Against Rotation Attacks

DOI: https://doi.org/10.1007/s11801-018-7212-0

Abstract-Summary

A robust digital watermarking algorithm is proposed based on the quaternion wavelet transform (QWT) and the discrete cosine transform (DCT) for copyright protection of color images.

The luminance component Y of a host color image in the YIQ space is decomposed by the QWT, and then the coefficients of four low-frequency subbands are transformed by a DCT.

The experimental results demonstrate that the proposed watermarking scheme shows strong robustness not only against common image processing attacks but also against arbitrary rotation attacks.

Acknowledgement
A machine generated summary based on the work of Han, Shao-cheng; Yang, Jinfeng; Wang, Rui; Jia, Gui-min
2018 in Optoelectronics Letters

3.1.4 Forensics Feature Analysis in Quaternion Wavelet Domain for Distinguishing Photographic Images and Computer Graphics

DOI: https://doi.org/10.1007/s11042-016-4153-0

Abstract-Summary
A novel set of features based on the Quaternion Wavelet Transform (QWT) is proposed for digital image forensics.

Compared with the Discrete Wavelet Transform (DWT) and the Contourlet Wavelet Transform (CWT), the QWT produces parameters, i.e., one magnitude and three angles, which provide more valuable information to distinguish photographic (PG) images and computer generated (CG) images.

It may be the first time to introduce the QWT to image forensics, but the improvements are encouraging.

Extended:

The QWT is applied to obtaining more feature information to achieve an improvement on the forensics scheme's identification performance.

A quaternion color wavelet transform for forensics will be studied.

Introduction
In [79], the statistical model built on the discrete wavelet transform (DWT) was proposed to obtain the relationship between the subband coefficients and color channels in 1999.

Chen and others [80] proposed a DWT- and DFT-based forensics scheme.

Özparlak and others [81] proposed to use the ridgelet and contourlet wavelet statistical model to obtain the regularity of the image on the basis of Farid's schemes.

We use the first four statistics extracted from the QWT domain as features for CG and PG classification.

In [44], a new multi-level Copula model is constructed based on the magnitude-phase dependance of the QWT, which obtained higher performance for texture classification than the DWT Copula.

(1) QWT as a new member of the wavelet family is applied to forensics; (2) The use of DWT and CWT is also investigated, and compared with the QWT.

Farid's Scheme
In [82], Farid proposed the linear predictor scheme to predict a new wavelet coefficient from the other wavelet coefficients in the same scale subband, the high-scale subband and the same scale subbands of the other two color channels.

The prediction error between the original wavelet coefficient and the predicted coefficient is generated and the first four order statistics of the prediction errors are used as features.

The oscillating wavelet filter is used to decompose the image that results in a complicated representation of a simple structure, including several coefficients in one neighborhood.

A small shift of the image causes substantial change in the energy of the wavelet transform, which increases the difficulty to extract valid features from wavelet coefficients.

Özparlak's Scheme

In [81], the author learned from the drawbacks of Farid's scheme and proposed a CWT-based statistical model to capture some features of images from the prediction error.

Compared with the DWT, the CWT overcomes some shortcomings and is more efficient in directionality and anisotropy, which catches smooth contours in different directions of an image.

In order to fit many directions, the CWT has some redundancy in its algorithmic aspect.

The Proposed QWT-based Scheme

The QWT is different from the DWT and the CWT, and it provides a local magnitude-phase analysis for images.

Due to the fact that a 2D H_T is equivalent to a 1D H_T along the x axis and/or y axis, a 1D Hilbert pair of wavelets and scaling functions are considered: In the QWT, the magnitude $|q|$ represents the signal strength at any space position in each frequency subband, similar to that of the DWT.

The first two QWT phases θ, υ indicate spatial shift information of image features in the axis x/y coordinate system, while the third phase φ indicates edge orientation mixtures and textual information.

Compared with the DWT and the CWT, the QWT includes not only the magnitude which encodes the frequency information but also three phases which indicate richer edge and texture information.

The dimension of the whole feature vector is 576, including magnitude features $4 \times 3 \times 3 \times 4 = 144$ for 4 statistics, 3 subbands, 3 scales and 4 complex planes and phase features $144 \times 3 = 432$ for 3 phases.

Experiments and Discussions

In order to obtain convincing experimental results and validly evaluate the performance of the proposed scheme, all the images with various sizes in the PG and CG database are used to calculate the statistical features for DWT-, CWT- and QWT-based schemes.

Three kinds of feature are tested, including the magnitude and phase features, the original and prediction features, and the four statistical features.

For the first kind of feature, its performance is tested by considering sub-features independently, including the magnitude feature, three phase features and combination features of the magnitude and three phases.

Besides, the accuracy of the phase feature excels that of the magnitude feature, which proves that the phase feature provides better performance of classification.

For the proposed scheme, the performance of the QWT feature excels the other two schemes, and achieves the highest forensics accuracy with feature dimension lower than Özparlak's scheme.

Conclusions and Future Work

The QWT is applied to obtaining more feature information to achieve the improvement on forensics scheme's identification performance.

The proposed scheme is constructed by generating the magnitude feature and the phase feature.

Various QWT features are analyzed and tested, including the magnitude and phase features, original and prediction features, and four statistical features, and compared with the features extracted in the DWT and the CWT domains.

The theoretical analysis and experimental results prove the proposed scheme's (based on QWT) superiority over the other schemes (based on DWT and CWT).

Acknowledgement

A machine generated summary based on the work of Wang, Jinwei; Li, Ting; Shi, Yun-Qing; Lian, Shiguo; Ye, Jingyu

2016 in Multimedia Tools and Applications

4 Medical Images and Tracking

Machine Generated Keywords QWT, fusion, method, medical, image fusion, frame, subband, sparse, video, low, source, sparse representation, resolution, motion, feature

4.1 Adopting Quaternion Wavelet Transform to Fuse Multi-Modal Medical Images

DOI: https://doi.org/10.1007/s40846-016-0200-6

Abstract-Summary

We propose a novel multi-modal medical image fusion method based on a simplified pulse-coupled neural network and the quaternion wavelet transform.

The proposed fusion algorithm is capable of combining not only pairs of computed tomography (CT) and magnetic resonance (MR) images, but also pairs of CT and proton-density-weighted MR images, and multi-spectral MR images such as T1 and T2.

Experiments on six pairs of multi-modal medical images are conducted to compare the proposed scheme with four existing methods.

It significantly outperforms existing medical image fusion methods in terms of subjective performance and objective evaluation metrics.

Introduction

A lot of medical image fusion methods based on multiscale geometry analysis have been proposed.

Examples include medical image fusion algorithms based on wavelets [83], the ripplet transform [84], the contourlet transform [85], the nonsubsampled contourlet transform (NSCT) [86], and the shearlet transform [87].

The quaternion wavelet transform (QWT) [16], proposed by Corrochano, is a new multiscale analysis tool for capturing the geometry features of an image.

Fusion rules based on principal component analysis methods lead to pixel distortion in fused multi-modal medical images [84].

In [88], a visibility feature method was proposed to fuse the quaternion wavelet coefficient of medical source images.

Although these fusion rules produce high-quality images, they also lead to loss of information and pixel distortion due to the nonlinear operations of fusion rules.

To overcome the aforementioned disadvantage, an improved medical image fusion model is proposed here based on the PCNN in the QWT domain.

Some experiments are performed to compare the proposed algorithm with other state-of-the-art medical image fusion methods.

Quaterion Wavelet Transform

The QWT is an extension of the complex wavelet transform that provides a richer scale-space analysis for 2-D signal geometric structure.

Compared with the traditional DWT, the QWT provides a magnitude-phases local analysis of images with a magnitude and three phases.

The other three DWTs obtained using the Hilbert transform can be applied to generate the three imaginary parts of the QWT.

Each subband of the QWT can be seen as an analytic signal associated with a narrow-band part of an image.

This QWT decomposition heavily depends on the position of the image with respect to the x and y axes (rotation-variance), and the wavelet is not isotropic.

Pulse-Coupled Neural Network

A PCNN is a visual-cortex-inspired neural network characterized by the global coupling and pulse synchronization of neurons.

A PCNN is a single-layer, 2-D, laterally connected neural network of pulse-coupled neurons.

A PCNN is a kind of feedback network that consists of many PCNN neurons.

The PCNN neuron model has three compartments: receptive field, modulation field, and pulse generator.

The neuron receptive area consists of the neighboring pixels of a corresponding pixel in the input image.

Proposed Scheme

MSD-based image fusion methods overcome these disadvantages because coefficients in subbands, not pixels or blocks in a spatial region, are considered as image details and selected as the coefficients of the fused image.

The key step in these approaches is to apply a multiscale transform to each source image to produce low- and high-frequency coefficients.

The inverse MSD is then applied to the fused coefficient to produce the fused medical image.

Different frequency coefficients can represent different features of source medical images.

The absolute-max-method is used to select the low-frequency coefficients and the PCNN-based method is used to fuse the high-frequency coefficients.

Most information in source images is in the low-frequency band.

In the absolute-max method, frequency coefficients from the source images with bigger absolute values are chosen as the fused coefficients.

The details of an image are mainly in the high-frequency coefficients.

Evaluation Metrics

Mutual information (MI), proposed by Piella [89], indicates how much information the fused image conveys about the reference image.

A larger MI indicates that the fused image contains more information from images A and B. Petrovic proposed an objective image fusion performance characterization [90] based on gradient information to evaluate fused medical images.

Experiments and Discussion

In Qu's method and Sudeb's scheme based on NSCT, the pyramid filter and the direction filter were set to 'pyrexc' and 'vk', respectively.

In Qu's method, the decomposition levels were set to [0,1,3,3,4].

In Sudeb's NSCT method, the decomposition levels were set [91, 92, 93] in accord with [94].

There is higher contrast in the fused images obtained using the proposed method and Sudeb's NSCT method than in those obtained using the other three methods.

These results show that the proposed method gives the most useful information in the fused images.

For these exceptions, the subjective effect of the proposed method is better than that of the other methods.

The proposed scheme is more effective than some state-of-the-art methods.

Conclusion

The proposed method adopts the PCNN fusion rule and abosolute-max fusion rule to select high- and low-frequency coefficients, respectively.

The experimental results illustrate that the proposed scheme is better than some existing fusion methods in terms of both objective and subjective metrics.

Acknowledgement
A machine generated summary based on the work of Geng, Peng; Sun, Xiuming; Liu, Jianhua
2017 in Journal of Medical and Biological Engineering

4.1.1 Correlative Feature Selection for Multimodal Medical Image Fusion Through QWT

DOI: https://doi.org/10.1007/978-3-030-00665-5_133

Abstract-Summary
A correlation-based feature selection is accomplished to extract an optimal feature set from the sub-bands thereby to reduce the computational time taken for the fusion process.

The QWT decomposes the source images first and then the feature selection process obtains an optimal feature set from the low-frequency (LF) as well as high-frequency (HF) sub-bands.

The obtained optimal feature sets of both LF sub-bands and HF sub-bands are fused through a low-frequency fusion rule and a high-frequency fusion rule and the fused LF and HF coefficients are processed through the IQWT to obtain a fused image.

Introduction
Medical image fusion is a compendious technique which gives a fused image as an input by fusing multiple images with different modalities.

The fused images are more suitable to help the doctor in the treatment planning: fusion of CT and MRI images can represent the bone structures and soft tissues in order to represent the physiological and anatomical features of a human body simultaneously.

In the case of spatial domain approaches, the fused image directly relates the pixel intensities of source image, whereas in the transform domain approaches, these are related with their transformed coefficients.

A novel image fusion technique is proposed in this paper to achieve an efficient and informative image by combining multiple medical images into one.

The source image is decomposed into sub-bands through the QWT first, and then the obtained bands are processed for correlation-based feature selection and then the selected coefficients are fused into their respective modified sub-bands.

Literature Survey
Kaur and others [95] proposed a PCA-based image fusion technique in collaboration with a genetic algorithm.

Further, based on the obtained regions, the source images are subjected to region segmentation and then the segmented regions are processed for final fusion according to their spatial frequencies with respect to the regions of the initial fused image.

The transform based image fusion approaches have gained a lot of research interest due to their effectiveness in the quality increment of fused images.

Compared to the DWT, the CT represents the details of image features more precisely by which the performance of fusion approach increases.

An effective DWT-based multi-focus image fusion approach is proposed by Yang and others [96] by seeing the physical meaning of wavelets.

Zhou and others [97] presented a novel multiscale fusion method based on a weighted gradient to solve the image misregistration problem which results in degradation of multi-focus images.

Preliminaries
Since the QWT results in phase-directed bands along with amplitude bands, a more detailed analysis can be acquired over the image with different phase variations.

From the 2D Hilbert transform, a 1D Hilbert transform can be derived through the x- and y-axis based on the above analytical extension, the 2D QWT can be defined, where the representations given the first three rows of the QWT equation [98, 99] are the mathematical formulae to extract the high-frequency coefficients from the image along diagonal, vertical, and horizontal directions, respectively.

In the case of the proposed QWT-based fusion framework, the QWT decomposes the image into low frequencies and high frequencies followed by magnitude and three phase bands.

The proposed correlation measure evaluation is applied over every band and a new feature set with maximum correlation is extracted for both source images A and B, then the extracted new and optimal feature set of both images are processed for fusion.

Fusion Framework
Consider the input images with two different models A and B. Apply the proposed fusion mechanism over these two images to obtain a fused image as follows: Apply the QWT over both source images A and B to decompose them into the low-frequency (Approximations, A) and high-frequency (Details, D) sub-band images.

Due to an issue of reduced contrast level in the fused image, this rule is violated and a novel fusion rule is derived considering the angular consistencies of approximation sub-bands.

The detailed sub-band image fusion is carried out through the most popular "Larger Absolute Selection Rule."

Due to the loss of vast complementary information in the fused image, this rule is not used here and a new fusion is derived based on the spatiofrequency energies of detailed sub-bands.

Simulation Results
To verify the performance of the proposed approach, an extensive simulation is carried out over various types of images like medical images, natural images, etc. Here, source images of size 256 * 256 are considered and the simulation is carried through MATLAB software.

BrainWeb: Simulated Brain Database: http://brainweb.bic.mni.mcgill.ca/brainweb/ Under the performance evaluation, five objective evaluation measurement parameters are adopted to evaluate the fusion performance.

There are weighted quality fusion index (WQFI) [100], local quality index (LQI) [100], edge-dependent fusion quality index (EFQI) [100], [101] which measures the transmission of edge features and visual features from input images to fused images, and mutual information (MI) [102] which measures the amount of information transferred from input images to output fused images.

Conclusion

A novel image fusion technique is proposed in this paper to achieve an efficient and informative image by combining multiple medical images into one.

This method accomplishes the QWT as a feature representation technique and correlation for feature selection.

Compared to the conventional multiscale transform techniques such as the DWT, CT, and NSCT, the QWT decomposes the image into the phase-deviated bands by which more information will be revealed about the frequency characteristics of an image.

Acknowledgement

A machine generated summary based on the work of Krishna Chaithanya, J.; Kumar, G. A. E. Satish; Ramasri, T.

2019 in Lecture Notes in Computational Vision and Biomechanics

4.1.2 Skin Lesion Image Classification Using Sparse Representation in Quaternion Wavelet Domain

DOI: https://doi.org/10.1007/s11760-021-02112-z

Abstract-Summary

Automated melanoma classification remains a challenging task because skin lesion images are prone to low contrast and many kinds of artifacts.

To handle these challenges, we introduce a novel and efficient method for skin lesion classification based on a machine learning approach and sparse representation (SR) in the quaternion wavelet (QW) domain.

Using the public skin lesion image datasets ISIC2017 and ISIC2019, we experimentally validated that creating a dictionary with quaternions of low-frequency wavelet sub-band leads to the most accurate classification of skin lesions to melanoma or benign.

Introduction

The large intra-class variation of melanomas in terms of color, texture, shape, size and location, in dermoscopy images, makes it difficult to discriminate melanomas from benign skin lesions.

To improve performance, our newly proposed supervised machine learning (ML) method takes advantages of Sparse Representation (SR) and the Quaternion Wavelet Transform (QWT) and does not use any lesion segmentation.

This paper proposes a novel supervised ML SR classification (SRC) method in the quaternion wavelet (QW) domain.

This study combines SR and different sub-bands of QW coefficients, to construct the dictionaries and classify in the QW domain.

We establish four quaternion sub-bands and classification is performed in each of them and every pair of them via two steps, namely sparse coding and label assignment by minimal residuals.

We name the novel method SRCQW, which stands for SRC in the QW domain.

Related Works
A survey of the automated skin lesion classification methods can be found in [103].

Paper [104] proposed classification based on the most clinically used ABCD (asymmetry, border irregularity, amount of colors and diameter) rule for melanoma detection.

Rastgoo and others [105] proposed a melanoma classification framework based on sparse coding without the pre-processing or lesion segmentation.

Codella and others [106] combined deep learning, sparse coding and SVM algorithms to better characterize lesions for melanoma classification.

The local texture and color features were extracted and then the SR of multiple features was jointly learned with a discriminative dictionary.

Moradi et al. [107] proposed a framework for melanoma segmentation and classification based on kernel SR.

Background
The main tool of the proposed SRCQW is quaternions [108].

The DWT decomposes an image into a low-frequency sub-band (LL) describing the image approximation information and three high-frequency sub-bands describing image details in horizontal (LH), vertical (HL) and diagonal (HH) directions, respectively [109].

Zou and others [110] showed, and [109] confirmed, that the use of the low-frequency sub-band alone provides the highest accuracy of face classification.

Classification with SR in QW Domain
To find the SR of the sample image, we apply the LASSO model [111] in the QW domain and call it QWLASSO.

To minimize the latter we modified the FISTA method [112] and made it work in the AQ.

The initial version of the QLASSO method was developed by Zou and others [113] where every quaternion represents the three color channels of an image.

Further, we develop the novel SRCQW method capable of working with every quaternion sub-band, or any subset of up to four quaternion sub-bands.

For the purpose of simplicity, hereafter, we call a vector/matrix of image approximation quaternions just quaternion vector/matrix.

Experimental Results

With ISIC2019, we use Monte Carlo cross-validation with c splits [114], and randomly select the training and test sets from the dataset in use.

To remedy the problem, we randomly select 626 malignant images from the HAM10000 [115] and ISIC Archive resource to obtain 1000 malignant images to be used for training.

Further, the proposed SRCQW was validated on the ISIC2019 dataset.

The Youden's index validates that the novel SRCQW method is best balanced in diagnosing both benign and melanoma images.

To validate the advantages of the proposed SRCQW over NNs, if middle to relatively small training sets are available, 10,000, 5000, and 1000 images from ISIC2019 have been randomly selected [116].

We randomly selected, 10 times, different sets of 1000 images from the training set of ISIC2019 to create the dictionary and used the rest for validation.

Conclusion and Discussion

Our primary contributions are in the development of the novel SRCQW method, which proved to be efficient for skin lesion classification in the AQ.

The SRCQW method is the first one that uses information from the wavelet domain to formulate the QWLASSO minimization problem for classification in the 4D space of the AQ.

SRCQW's drawback is its relatively high computational complexity, which requires the use of a portion of the training set, if it is too large.

Using only 876 training images, we classify the set of 10,982 test images with a score of 0.7804, according to the ISIC2020 challenge's metric.

The methods, which conducted experiments with ISIC2020 and received higher results than ours, used the entire set of 33,126 training images.

To let the SRCQW method use a large training dataset, a significant computer resource is needed.

Acknowledgement

A machine generated summary based on the work of Ngo, Long H.; Luong, Marie; Sirakov, Nikolay M.; Viennet, Emmanuel; Le-Tien, Thuong

2022 in Signal, Image and Video Processing

4.1.3 A Fast Quaternion Wavelet-Based Motion Compensated Frame Rate Up-Conversion with Fuzzy Smoothing: Application to Echocardiography Temporal Enhancement

DOI: https://doi.org/10.1007/s11042-020-09834-1

Abstract-Summary

We propose a fast Frame Rate Up-Conversion (FRUC) method based on Quaternion Wavelet Transform (QWT) motion estimation to improve the motion estimation accuracy and reduce the computational complexity.

The proposed method contains three key elements: motion estimation, motion post-processing, and motion compensated frame interpolation.

The evaluations confirm that while our proposed method keeps the PSNR and SSIM performance suitable, it is at least twice faster in comparison with reference methods on the benchmark dataset.

The echocardiography evaluations reveal our proposed method's superiority in terms of PSNR, SSIM, and computation time, which make the proposed method suitable as an echocardiographic-specific method.

Extended:

We have proposed a fast QWT based frame rate up-conversion method with fuzzy smoothing to do motion-compensated frame interpolation.

Introduction

FRUC methods are among the oldest methods of temporal video enhancement and applied in a broad category of video processing applications like format conversion (for example from 24 frames per second film content to 30 frames per second video content), low bit rate video coding, slow-motion playback, and rate allocation policy [117].

The second class of FRUC methods, which is called Motion-Compensated FRUC(MC-FRUC [118, 119, 120, 121–122, 123, 124, 125, 126, 127, 128, 129, 130, 131, 117, 132–133], are based on Motion Estimation (ME) to provide superior quality high frame rate video.

Due to computational efficiency, block-based motion estimation methods are common choices for FRUCs.

Tai and others [134] proposed a robust multi-pass true motion estimation algorithm to enhance the accuracy of the motion vector fields in frame rate up-conversion applications.

Phased based motion estimation methods are among the robust ME algorithms, which have been used in the field of video motion processing.

Proposed Method

The 1D dual-tree complex wavelet transform (CWT) expands a real signal using the wavelet function $\psi_c(t) = \psi_h(t) + j\psi_g(t)$ in which $\psi_h(t)$ and $\psi_g(t)$ are obtained from two filter banks, that play the role of the real and imaginary parts of a complex analytic wavelet (the imaginary part is the Hilbert transform of the real wavelet ($\psi_h(t)$).).

The QWT consists of a standard DWT tensor wavelet and three 1D Hilbert transformations of a real wavelet along x, y, or both coordinates.

The fuzzy motion vector smoothing consists of two steps, MV smoothness calculation, which is a measure of MV outlierness degree, and weighted Vector Mean Filtering (VMF).

To cope with this problem, we calculate the degree of outlierness (1 − smoothnessdegree) for each MV in the first step, then based on the smoothness degree of neighboring MVs, the filter output will be the result of weighted vector mean filtering with smoothness degree as the MV weights.

Experiments and Results

To evaluate our proposed method performance on B-mode echocardiography, we have used STRAUS, REVSUS (Realistic Vendor-Specific Synthetic Ultrasound Data), and CAMUS datasets, which are benchmark datasets in echocardiography image processing.

We have also compared our proposed method with reference algorithms in terms of mean computation time for each frame to reconstruct in sec.

Since PSNR and SSIM measures cannot accurately represent the perceived image quality by the human visual system, there is also a need for a subjective evaluation by comparing the original frames with the ones interpolated with proposed and reference algorithms.

We have also assessed the performance of our proposed method with reference ones on echocardiography sequences obtained from the STRAUS dataset in the second part of our evaluation.

In the third part of our evaluation, we evaluate the performance of our proposed method and TSRBS on REVSUS and CAMUS datasets.

Discussion

This paper has introduced a frame rate up-conversion method with a phase-based low complexity QWT motion estimation and fuzzy smoothing.

It is composed of QWT motion estimation, fuzzy motion vector smoothing, bilateral motion refinement, and adaptive overlapped block motion compensation.

In the proposed method, we have estimated the motion vector field, based on the phase difference in the QWT of the previous and current frames, which have not been used in FRUC applications before.

QWT motion estimation is a phase-based low complexity motion estimation method.

To vector median and vector mean filters which are the typical choices for smoothing, a fuzzy motion vector smoothing filter assumes that all MVs in a neighborhood have some degrees of outlierness which should be considered when doing smoothing.

Conclusion

We evaluate the proposed method on the YUV benchmark video sequences as well as three simulated and realistic echocardiography datasets both qualitatively and quantitatively.

The results on the YUV video dataset demonstrated that our proposed method performance is near to high-performance reference methods in terms of PSNR and SSIM.

The proposed method is the fastest method according to the computation time results (the computation time of our proposed method in comparison with references decreased by a factor 1.61 to 49.82).

The results show that our proposed method improves the PSNR by 10.17%, 4.72%, and 7.06% and the SSIM by 17.81%, 6.85%, and 5.40% in STRAUS, REVSUS, and CAMUS respectively in comparison with the TSRBS (which is the best echocardiographic-specific reference method in our evaluations).

The evaluations on echocardiography datasets reveal that our proposed method could be used as an echocardiographic-specific method for temporal enhancement.

Acknowledgement
A machine generated summary based on the work of Khoubani, Sahar; Moradi, Mohammad Hassan
2020 in Multimedia Tools and Applications

4.1.4 Visual Objects Tracking and Identification Based on Reduced Quaternion Wavelet Transform

DOI: https://doi.org/10.1007/s11760-014-0636-5

Abstract-Summary
A new method for tracking moving objects based on a reduced a quaternion wavelet transform by using color space information is proposed.

The reduced quaternion wavelet transform is a new multi-scale analysis tool for geometric image features.

The new proposed method can calculate the mean value and edge energy of visual objects by using the reduced quaternion wavelet coefficients.

Extended:

We apply the RQWT to the region of moving objects after making a color space transformation which is from RGB color space to HIS color space.

The reduced quaternion wavelet transform (RQWT) uses wavelets which are providing reduced quaternion valued sub-bands with shift-invariant magnitude and three new RQWT phases.

Experimental results show that the proposed method can obtain better identification rate in comparison with some previous methods.

Introduction
With the development of the technology, moving object tracking is one of the most important problems in image processing and machine vision.

Several moving object tracking algorithms have been proposed.

Kuo and others [135] proposed a new appearance model for moving object tracking, which allows the model to fit the appearance variation.

Zhou and others [136] proposed the SIFT feature in moving object tracking.

Multiple instances learning (MIL) was applied to solve the drifting problem in moving object tracking [137].

Cetin and others [138] proposed a method and system for characterizing the motion of moving objects in a video.

Toreyin and others [139, 140] applied the DWT for moving object and region detection in a video.

Cheng and others [141] proposed a new method based on the DWT for tracking multiple moving objects.

Reduced Quaternion Wavelet Transform

Schute and Wenzel [142] introduced the reduced quaternions, in which multiplication rules are more commutative compared with standard quaternions.

The reduced quaternionic multiplication is commutative.

It is considered as a local reduced quaternion Fourier transform (RQFT).

The 2D RQWT is defined as follows: Note that the magnitude describes the local geometric information of the original signal, and three phases indicate the 2D structure features of the signal.

Moving Object Tracking Based on RQWT

The moving objects are calculated by extracting connected components.

The process to track moving objects is to find the region of objects from the difference image.

After locating the moving objects in the sequence, we can obtain the boundary of the moving objects.

We can compute the color information of the moving pixels with their neighbors in the sequence.

It is an important statistical characteristic of the moving object.

Experimental Results

Several experiments are conducted to prove the effectiveness of the proposed moving tracking method.

The second experiment is to test the effectiveness of the proposed algorithm by using several sequences.

The third experiment is conducted to evaluate the recognition rate for different sequences.

The fourth experiment is used to illustrate the robustness and effectiveness of the proposed method compared with other methods.

It is clear to see that the proposed object tracking method has a sharp peak and fast computation speed.

The common problem of background culling is solved by the proposed method more effectively than the other object tracking methods.

Conclusion

A new method for detecting and tracking moving objects based on the reduced quaternion wavelet transform is proposed in this paper.

The statistical characteristics of mean value and standard deviation are used as feature for object recognition.

Experimental results show that the proposed method can obtain better identification rate in comparison with some previous methods.

Acknowledgement

A machine generated summary based on the work of Gai, Shan; Luo, Limin 2014 in Signal, Image and Video Processing

References

1. Hitzer, E.: Clifford (geometric) algebra wavelet transform. In: Skala, V., Hildenbrand, D. (eds.), Roc. of GraVisMa 2009, 02–04 September 2009, Plzen, Czech Republic, pp. 94–101 (2009)
2. Bahri, M., Ashino, R., Vaillancourt, R.: Two-dimensional quaternion wavelet transform. Appl. Math. Comput. 218, 10–21 (2011)
3. Heisenberg, W.: Uber den anschaulichen inhalt der quanten theoretischen kinematik und mechanik. Zeitschrift für Physik 43, 172–198 (1927)
4. Chen L-P, Kou K.-I., Liu M.-S.: Pitt's inequality and the uncertainty principle associated with the quaternion Fourier transform. J. Math. Anal. Appl. 423, 681–700 (2015)
5. Kou, K.I., Xu, R.H.: Windowed linear canonical transform and its applications. Signal Process. 92(1), 179–188 (2012)
6. Wang, J., Wang, Y., et al.: Discrete linear canonical wavelet transform and its applications. EURASIP J. Adv. Signal Process. 29, (2018). https://doi.org/10.1186/s13634-018-0550-z
7. Wei, D., Li, Y.: Generalized wavelet transform based on the convolution operator in the linear canonical transform domain. Optik. 125(16), 4491–4496 (2014)
8. Shah, F.A., Tantary, A.Y.: Linear canonical Stockwell transform. J. Math. Anal. Appl. 443, 1–28 (2020)
9. Guo, Y., Li, B.Z.: The linear canonical wavelet transform on some function spaces. Int. J. Wavelets Multiresolut. Inf. Process. 16(1), 1850010 (2018). https://doi.org/10.1142/S0219691318500108
10. Shah, F.A., Teali, A.A., Tantary, A.Y.: Special affine wavelet transform and the corresponding Poisson summation formula. Int. J. Wavelets Multiresolut. Inf. Process. (2021). https://doi.org/10.1142/S0219691320500861
11. Bahri, M., Ashino, R., Vaillancourt, R.: Continuous quaternion Fourier and wavelet transforms. Int. J. Wavelets Multiresolut. Inf. Process. 12, 1460003 (2014)
12. Akila, L., Roopkumar, R.: Ridgelet transform on quaternion valued functions. Int. J. Wavelets Multiresolut. Inf. Process. 14, 1650006 (2016)
13. Akila, L., Roopkumar, R.: Quaternionic Stockwell transform. Integ. Transf. Spec. Funct. 27(6), 484–504 (2016)
14. Akila, L., Roopkumar, R.: Quaternionic curvelet transform. Optik. 131, 255–266 (2017)
15. Shah, F.A., Tantary, A.Y.: Quaternionic Shearlet transform. Optik. 175, 115–125 (2018)
16. Bayro-Corrochano E.: The theory and use of the quaternion wavelet transform. J. Math. Imaging Vis. 24, 19–35 (2006)
17. Gao, W.B., Li, B.Z.: Quaternion windowed linear canonical transform of two-dimensional signals. Adv. Appl. Clifford Algebras. 30, (2020). https://doi.org/10.1007/s00006-020-1042-4
18. Folland, G.B., Sitaram, A.: The uncertainty principle: a mathematical survey. J. Fourier Anal. Appl. 3(3), 207–238 (1997)
19. Wilczok, E.: New uncertainty principles for the continuous Gabor transform and the continuous wavelet transform. Doc. Math. 5, 201–226 (2000)
20. M. Bahri, E.S.M. Hitzer, A. Hayashi, R. Ashino, An uncertainty principle for quaternion Fourier transform. Comput. Math. Appl. 56 (9), 2398–2410 (2008)
21. Ell, T.A., Bihan, N.L., Sangwine, S.J.: Quaternion Fourier Transforms for Signal and Image Processing. Wiley, New York (2014)
22. Hitzer, E.: Quaternion Fourier transform on quaternion fields and generalizations. Adv. Appl. Clifford Algebra 17, 497–517 (2007). https://doi.org/10.1007/s00006-007-0037-8, preprint: http://arxiv.org/abs/1306.1023
23. Hitzer, E.: Directional uncertainty principle for quaternion Fourier transforms. Adv. Appl. Clifford Algebra 20(2), 271–284 (2010). https://doi.org/10.1007/s00006-009-0175-2, preprint: http://arxiv.org/abs/1306.1276
24. Bülow, T.: Hypercomplex spectral signal representations for the processing and analysis of images, Ph.D. Thesis, University of Kiel 9903, 161 pages (1999)

25. Bülow, T., Felsberg, M., Sommer, G.: Non-commutative hypercomplex Fourier transforms of multidimensional signals. In: Sommer, G. (ed.) Geometric Computing with Clifford Algebras. Theor. Found. and Appl. in Comp. Vision and Robotics, pp. 187–207. Springer, Berlin (2001)
26. Antoine, J.P., Murenzi, R.: Two-dimensional directional wavelets and the scale-angle representation. Sig. Process. 52(3), 259–281 (1996)
27. Antoine, J.P., Vandergheynst, P., Murenzi, R.: Two-dimensional directional wavelets in image processing. Int. J. Imag. Syst. Technol. 7(3), 152–165 (1996)
28. Bahri, M.: Quaternion algebra-valued wavelet transform. Appl. Math. Sci. 5(71), 3531–3540 (2011)
29. Hitzer, E., Sangwine, S.J. (eds.): Quaternion and Clifford–Fourier Transforms and Wavelets, Trends in Mathematics, pp. 57–83, Springer, Basel (2013)
30. Achak, A., Bouhlal, A., Daher, R., Safouane, N.: Titchmarsh's theorem and some remarks concerning the right-sided quaternion Fourier transform. Boletín de la Sociedad Matemática Mexicana (2020)
31. Bouhlal, A., Achak, A., Daher, R., Safouane, N.: Dini–Lipschitz functions for the quaternion linear canonical transform. Rend. Circ. Mat. Palermo Ser. 2 (2020)
32. Potapov, M.K.: Application of the operator of generalized translation in approximation theory. Vestnik Moskovskogo Universiteta, Seriya Matematika, Mekhanika 3, 38–48 (1998)
33. Platonov, S.S.: Generalized Bessel Translations and Certain Problems of the Theory of Approximation of Functions in the Metrics of $L_{2,\alpha}$. I (2000)
34. Bayro-Corrochano, E.: Geometric Computing for Wavelet Transforms, Robot Vision, Learning, Control and Action. Springer, Berlin (2010)
35. Kovesi, P.: Invariant measures of images features from phase information. PhD thesis, University of Western Australia (1996)
36. Kravchenko, V.F., Perez-Meana, H.M., Ponomaryov, V.I.: Adaptive Digital Processing of Multidimensional Signals with Applications. Fizmatlit, Moscow (2009)
37. Guyaev, Yu.V., Kravchenko, V.F.: A new class of WA-systems of Kravchenko–Rvachev functions. Moscow Dokl. Math. 75(2), 325–332 (2007)
38. Gorshkov, A., Kravchenko, V.F., Rvachev, V.A.: Estimation of the discrete derivative of a signal on the basis of atomic functions. Izmer. Tekh. 1(8), 10 (1992)
39. Rvachev, V.A.: Compactly supported solution of functional–differential equations and their applications. Russ. Math. Surv. 45(1), 87–120 (1990)
40. Hamilton, W.R.: Elements of Quaternions. Chelsea, New York (1969). Longmans Green, London (1866)
41. Delanghe, R.: Clifford analysis: history and perspective. In: Computational Methods and Function Theory, vol. 1, pp. 107–153 (2001)
42. Radon, J.: Über die Bestimmung von Funktionen durch ihre Integralwerte längs gewisser Mannigfaltigkeiten. Ber. Sächs. Akad. Wiss., Leipzig. Math.-Phys. Kl. 69, 262–277 (1917)
43. Gai, S., Yang, G., Wan, M., Wang, L.: Hidden Markov tree model of images using quaternion wavelet transform. Comput. Electr. Eng. 40(3), 819–832 (2014)
44. Liu, C., Li, J., Fu, B.: Magnitude-phase of quaternion wavelet transform for texture representation using multilevel copula. IEEE Sig. Process. Lett. 20(8), 799–802 (2013)
45. Chan, W.L., Choi, H., Baraniuk, R.G.: Coherent multiscale image processing using dual-tree quaternion wavelets. IEEE Trans. Image Process. 17(7), 1069–1082 (2008)
46. Zhang X, Feng X (2014) Multiple-step local Wiener filter with proper stopping in wavelet domain. J Vis Commun Image Represent 25(2):254–262
47. Bayrocorrochano, E.: Multiresolution image analysis using the quaternion wavelet transform. Numer. Algorithms 39(1–3), 35–55 (2005)
48. Shan G, Liu P, Liu J, Tang X (2010) A new image denoising algorithm via bivariate shrinkage based on quaternion wavelet transform. J Comput Inf Sys 6(11):3751–3760
49. Luisier F, Blu T, Unser M (2007) A new SURE approach to image denoising: inter-scale orthonormal wavelet thresholding. IEEE Trans Image Process 16(3):593–606

50. Miller M, Kingsbury N (2008) Image denoising using derotated complex wavelet coefficients. IEEE Trans Image Process 17(9):1500–1511
51. Po D, Do MN (2006) Directional multiscale modeling of images using the contourlet transform. IEEE Trans Image Process 15(6):1610–1620
52. Cunha AL, Jianping Z, Do MN (2006) The nonsubsampled contourlet transform: theory, design, and applications. IEEE Trans Image Process 15(10):3089–3101
53. Mojzis, F., Svihlik, J., Fliegel, K., Knazovicka, L., Jerhotova, E.: Measurement and Analysis of Real Imaging Systems, Radioengineering, Vol. 21, No. 1, April (2012).
54. Kingsbury, N.G.: The dual-tree complex wavelet transform: a new efficient tool for image restoration and enhancement, in the 9th European Signal Processing Conference (EUSIPCO), pp. 319–322, (1998).
55. Kingsbury, N.G.: A dual-tree complex wavelet transform with improved orthogonality and symmetry properties, IEEE International Conference on Image Processing(ICIP), pp. 375–378, (2000).
56. Selesnick, I., W., Baraniuk, R., G., Kingsbury, N., G.: A coherent framework for multiscale signal and image processing, IEEE Signal Processing Magazine, November (2005).
57. Kumar, S., Kumar. S., Sukavanam, N., Raman, B.: Dual tree fractional quaternion wavelet transform for disparity estimation, ISA Transactions, Elsevier, 547–559, (2014).
58. Jansen, M.: Noise Reduction by Wavelet Thresholding, Lecture notes in statistics (ISSN 0930–0325 ; 161), Springer, (2001).
59. Karybali, I.G., Berberidis, K.: Efficient spatial image watermarking via new perceptual masking and blind detection schemes. IEEE Trans. Inf. Forensics Secur. 1(2), 256–274 (2006)
60. Nasir, I., Weng, Y., Jiang, J., Ipson, S.: Multiple spatial watermarking technique in color images. SIViP 4(2), 145–154 (2010)
61. Das, C., Panigrahi, S., Sharma, V.K., Mahapatra, K.K.: A novel blind robust image watermarking in DCT domain using inter-block coefficient correlation. AEU – Int. J. Electron. Commun. 68(3), 244–253 (2014)
62. Urvoy, M., Goudia, D., Autrusseau, F.: Perceptual DFT, watermarking with improved detection and robustness to geometrical distortions. IEEE Trans. Inf. Forensics Secur. 9(7), 1108–1119 (2014)
63. Keyvanpour, M.R., Bayat, F.M.: Blind image watermarking method based on chaotic key and dynamic coefficient quantization in the DWT domain. Math. Comput. Model. 58(1–2), 56–67 (2013)
64. Mardanpour, M., Chahooki, M.A.Z.: Robust transparent image watermarking with Shearlet transform and bidiagonal singular value decomposition. AEU – Int. J. Electron. Commun. 70(6), 790–798 (2016)
65. Chu, W.C.: DCT-based image watermarking using subsampling. IEEE Trans. Multimedia 5(1), 34–38 (2003)
66. Yin, M., Liu, W., Shui, J., Wu, J.: Quaternion wavelet analysis and application in image denoising. Math. Probl. Eng. 2012(1), 587–612 (2012)
67. Lu, W., Lu, H., Chung, F.: Robust digital image watermarking based on subsampling. Appl. Math. Comput. 181(2), 886–893 (2006)
68. Su, Q., Niu, Y., Wang, Q., Sheng, G.: A blind color image watermarking based on DC component in the spatial domain. Optik – Int. J. Light Electron Opt. 124(23), 6255–6260 (2013)
69. Liu, N., Li, H., Dai, H., Chen, D.: Robust blind image watermarking based on chaotic mixtures. Nonlinear Dyn. 80(3), 1329–1355 (2015)
70. Mainali, P. (2014). Robust registration in integrated hyperspectral imaging (robuuste registratie in geïntegreerde, hyperspectrale beeldopname). 2014. Web.
71. Yang, H. Y., Wang, P., Wang, X. Y., et al. (2015). Robust digital watermarking based on local invariant radial harmonic fourier moments. Multimedia Tools and Applications, 74(23), 10559–10579.
72. Fletcher, P., Sangwine, S.J.: The development of the quaternion wavelet transform. Sig. Process. 136, 2–15 (2017). https://doi.org/10.1016/j.sigpro.2016.12.025

73. Mainali, P., Lafruit, G., Yang, Q., et al. (2013). SIFER: Scale-invariant feature detector with error resilience. International Journal of Computer Vision, 104(2), 172–197.
74. Aslantas, V., Ozer, S., Ozturk, S.: Improving the performance of DCT-based fragile watermarking using intelligent optimization algorithms. Opt. Commun. 282(14), 2806–2817 (2009)
75. Ali, M., Ahn, C.W.: An optimized watermarking technique based on self-adaptive de in DWT-SVD transform domain. Signal Process. 94(1), 545–556 (2014)
76. Lei, B., Tan, E.-L., Chen, S., Ni, D., Wang, T., Lei, H.: Reversible watermarking scheme for medical image based on differential evolution. Expert Syst. Appl. 41(7), 3178–3188 (2014)
77. Lei, B., Soon, I.Y., Tan, E.L.: Robust SVD-based audio watermarking scheme with differential evolution optimization. IEEE Trans. Audio Speech Lang. Process. 21(11), 2368–2378 (2013)
78. Jamal, S.S., Shah, T., Hussain, I.: An efficient scheme for digital watermarking using chaotic map. Nonlinear Dyn. 73(3), 1469–1474 (2013)
79. Buccigrossi R, Simoncelli E (1999) Image compression via joint statistical characterization in the wavelet domain. IEEE Trans Image Process 8(12):1688–701
80. Chen W, Shi Y, Xuan G (2007) Identifying computer graphics using HSV color model and statistical moments of characteristic functions. In: Proceedings of ICME, pp. 1123–1126
81. Özparlak L, Avcıbaş I (2011) Differentiating between images using wavelet-based transforms: a comparative study. IEEE Trans Inf Forensics Secur 6(4):1418–1431
82. Lyu S, Farid H (2005) How realistic is photorealistic? IEEE Trans Signal Process 53(2):845–850
83. Zhang, X., Zheng, Y., Peng, Y., & Liu, W. (2009). Research on multi-mode medical image fusion algorithm based on wavelet transform and the edge characteristics of images. International Congress on Image and Signal Processing, 1–4.
84. Das, S., & Kundu, M. K. (2011). Ripplet based multimodality medical image fusion using pulse-coupled neural network and modified spatial frequency. International Conference on Recent Trends in Information Systems, 229–234.
85. Rajkumar, S., & Kavitha, S. (2010). Redundancy discrete wavelet transform and contourlet transform for multimodality medical image fusion with quantitative analysis. International Conference on Emerging Trends in Engineering and Technology, 134–139. IEEE Computer Society.
86. Bhatnagar, G., Wu, Q. M. J., & Liu, Z. (2013). Directive contrast based multimodal medical image fusion in NSCT domain. IEEE Transactions on Multimedia, 15(5), 1014–1024.
87. Wang, Z. (2012). Image fusion by pulse couple neural network with shearlet. Optical Engineering, 51(6), 067005.
88. Peng, G., Xing, S., & Tan, X. (2015). Medical image fusion based on quaternion wavelet transform and visibility feature. International Journal of Applied Mathematics and Machine Learning, 2(1), 9–26.
89. Piella, G. (2002). A general framework for multiresolution image fusion: From pixels to regions. Information Fusion, 4(4), 259–280.
90. Petrovic, V., & Xydeas, C. (2005). Objective image fusion performance characterisation. Tenth IEEE International Conference on Computer Vision, 2, 1866–1871.
91. Kettenbach, J., Wong, T. D., Hata, N., Schwartz, R., Black, P., Kikinis, R., et al. (1999). Computer-based imaging and interventional MRI: Applications for neurosurgery. Computerized Medical Imaging and Graphics, 23(5), 245–258.
92. Zhu, Y. M., & Cochoff, S. M. (2006). An object-oriented framework for medical image registration, fusion, and visualization. Computer Methods and Programs in Biomedicine, 82(3), 258–267.
93. Petrovic, V. S., & Xydeas, C. S. (2004). Gradient-based multiresolution image fusion. IEEE Transactions on Image Processing, 13(2), 228–237.
94. Das, S., & Kundu, M. K. (2012). NSCT-based multimodal medical image fusion using pulse-coupled neural network and modified spatial frequency. Medical & Biological Engineering & Computing, 50(10), 1105–1114.

95. Kaur R (2016) An approach for image fusion using PCA and genetic algorithm. Int J Comput Appl 145(6):(0975–8887)
96. Yang Y, Huang SY, Gao J, Qian Z (2014) Multi-focus image fusion using an effective discrete wavelet transform based algorithm. Meas Sci Rev 14(2):102–108
97. Zhou Z, Li S, Wang B (2014) Multi-scale weighted gradient-based fusion for multi-focus images. Inf Fusion 20:60–67
98. Yin M, Liu W, Shui J, Wu J (2012) Quaternion wavelet analysis and application in image denoising. Math Prob Eng 2012, Art. no. 493976
99. Gai S, Wang L, Yang G, Yang P (2016) Sparse representation based on vector extension of reduced quaternion matrix for multiscale image denoising. IET Image Process 10(8):598–607
100. Piella G, Heijmans H (2003) A new quality metric for image fusion. In: Proceedings of the international conference on image processing (ICIP '03), Catalonia, Spain, pp 173–176, September 2003
101. Xydeas CS, Petrović V (2000) Objective image fusion performance measure. Electron Lett 36(4):308–309
102. Qu G, Zhang D, Yan P (2002) Information measure for performance of image fusion. Electron Lett 38(7):313–315
103. Oliveira, R.B., Papa, J.P., Pereira, A.S., Tavares, J.M.R.: Computational methods for pigmented skin lesion classification in images: review and future trends. Neural Comput. Appl. 29(3), 613–636 (2018)
104. Sirakov, N.M., Mete, M., Selvaggi, R., Luong, M.: New accurate automated melanoma diagnosing systems. In: 2015 ICHI, pp. 374–379. IEEE (2015)
105. Rastgoo, M., Lemaître, G., et al.: Classification of melanoma lesions using sparse coded features and random forests. In: Medical Imaging 2016: CAD. Int. Society for Optics and Photonics (2016)
106. Codella, N., Cai, J., Abedini, M., Garnavi, R., Halpern, A., Smith, J.R.: Deep learning, sparse coding, and SVM for melanoma recognition in dermoscopy images. In: Int. Workshop on Machine Learning in Medical Imaging, pp. 118–126. Springer (2015)
107. Moradi, N., Mahdavi-Amiri, N.: Kernel sparse representation based model for skin lesions segmentation and classification. Comput. Methods Progr. Biomed. 182, 105038 (2019)
108. Girard, P.: Quaternions, Clifford Algebras and Relativistic Physics. Springer, Berlin (2007)
109. Ngo, L.H., Luong, M., Sirakov, N.M., Le-Tien, T., Guerif, S., Viennet, E.: Sparse representation wavelet based classification. In: 25th IEEE ICIP, pp. 2974–2978 (2018)
110. Zou, W., Li, Y.: Image classification using wavelet coefficients in low-pass bands. In: IEEE Int. Joint Conf. on Neural Networks, pp. 114–118 (2007)
111. Tibshirani, R.: Regression shrinkage and selection via the lasso. J. R. Stat. Soc. Ser. B (Methodol.) 58(1), 267–288 (1996)
112. Beck, A., Teboulle, M.: A fast iterative shrinkage-thresholding algorithm for linear inverse problems. SIIMS 2(1), 183–202 (2009)
113. Zou, C., Kou, K.I., Wang, Y.: Quaternion collaborative and sparse representation with application to color face recognition. IEEE TIP 25(7), 3287–3302 (2016)
114. Dubitzky, W., Granzow, M., Berrar, D.P.: Fundamentals of Data Mining in Genomics and Proteomics. Springer, Berlin (2007)
115. Tschandl, P., Rosendahl, C., Kittler, H.: The ham10000 dataset, a large collection of multi-source dermatoscopic images of common pigmented skin lesions. Sci. Data 5, 180161 (2018)
116. Pollastri, F., Parreño, M., Maroñas, J., Bolelli, F., Paredes, R., Ramos, D., Grana, C.: A deep analysis on high resolution dermoscopic image classification. IET Res. J. (2021)
117. Zhang Y, Zhao D, Ji X, Wang R, Gao W (2009) A spatio-temporal auto regressive model for frame rate upconversion. IEEE Trans Circ Syst Video Technol 19(9):1289–1301. https://doi.org/10.1109/TCSVT.2009.2022798
118. Chen G (2020) Frame rate up-conversion algorithm based on adaptive-agent motion compensation combined with semantic feature analysis. J Ambient Intell Human Comput 11(2):511–518. https://doi.org/10.1007/s12652-018-0974-1

119. Chen T (2002) Adaptive temporal interpolation using bidirectional motion estimation and compensation. IEEE Int Conf Image Process 2:393–396. https://doi.org/10.1109/icip.2002.1039950
120. Choi BD, Han JW, Kim CS, Ko SJ (2007) Motion-compensated frame interpolation using bilateral motion estimation and adaptive overlapped block motion compensation. IEEE Trans Circ Syst Video Technol 17(4):407–415. https://doi.org/10.1109/TCSVT.2007.893835
121. Choi BT, Lee SH, Ko SJ (2000) New frame rate up-conversion using bi-directional motion estimation. IEEE Trans Consum Electron 46 (3):603–609. https://doi.org/10.1109/30.883418
122. Choi G, Heo PG, Park HW (2019) Triple-frame-based bi-directional motion estimation for motion-compensated frame interpolation. IEEE Trans Circ Syst Video Technol 29(5):1251–1258. https://doi.org/10.1109/TCSVT.2018.2840842
123. Kang SJ, Yoo S, Kim YH (2010) Dual motion estimation for frame rate up-conversion. IEEE Trans Circ Syst Video Technol 20(12):1909–1914. https://doi.org/10.1109/TCSVT.2010.2087832
124. Kim US, Sunwoo MH (2014) New frame rate up-conversion algorithms with low computational complexity. IEEE Trans Circ Syst Video Technol 24 (3):384–393. https://doi.org/10.1109/TCSVT.2013.2278142
125. Lee SH, Kwon O, Park RH (2003) Weighted-adaptive motion-compensated frame rate up-conversion. IEEE Trans Consum Electron 49(3):485–492. https://doi.org/10.1109/TCE.2003.1233759
126. Li R, Liu H, Chen J, Gan Z (2016) Wavelet pyramid based multi-resolution bilateral motion estimation for frame rate up-conversion. IEICE Transactions on Information and Systems E99D(1):208–218, https://doi.org/10.1587/transinf.2015EDP7027
127. Li R, Ji B, Li Y, Wu C (2019) A Bayer motion estimation for motion-compensated frame interpolation. Multimed Tools Appl 78 (14):19603–19619. https://doi.org/10.1007/s11042-019-7337-6
128. Van Thang N, Choi J, Hong JH, Kim JS, Lee HJ (2018) Hierarchical motion estimation for small objects in frame-rate up-conversion. IEEE Access 6:60353–60360. https://doi.org/10.1109/ACCESS.2018.2875688
129. Wang D, Vincent A, Blanchfield P, Klepko R (2010) Motion-compensated frame rate up-conversion-Part II: new algorithms for frame interpolation. IEEE Trans Broadcast 56(2):142–149. https://doi.org/10.1109/TBC.2010.2043895
130. Yoon SJ, Kim HH, Kim M (2018) Hierarchical extended bilateral motion estimation-based frame rate upconversion using learning-based linear mapping. IEEE Trans Image Process 27(12):5918–5932. https://doi.org/10.1109/TIP.2018.2861567
131. Zhai J, Yu K, Li J, Li S (2005) A low complexity motion compensated frame interpolation method. Proceedings – IEEE International Symposium on Circuits and Systems pp 4927–4930, https://doi.org/10.1109/ISCAS.2005.1465738
132. Zhang Y, Chen L, Yan C, Qin P, Ji X, Dai Q (2020) Weighted convolutional motion-compensated frame rate up-conversion using deep residual network. IEEE Trans Circ Syst Video Technol 30(1):11–22. https://doi.org/10.1109/TCSVT.2018.2885564
133. Zhou L, Sun R, Tian X, Chen Y (2018) Phase-based frame rate up-conversion for depth video. J Electron Imaging 27(04):1. https://doi.org/10.1117/1.jei.27.4.043036
134. Tai SC, Chen YR, Huang ZB, Wang CC (2008) A multi-pass true motion estimation scheme with motion vector propagation for frame rate up-conversion applications. IEEE/OSA J Display Technol 4 (2):188–197. https://doi.org/10.1109/JDT.2007.916014
135. Kuo, C., Huang, C., Nevatia, R.: Multi-target tracking by on-line learned discriminative appearance models. IEEE Int. Conf. Comput. Vis. Pattern Recognit. 8, 1–8 (2010)
136. Zhou, H., Yuan, Y., Shi, C.: Object tracking using SIFT features and mean shift. J. Comput. Vis. Image Underst. 113(3), 345–352 (2009)
137. Babenko, B., Yang, M., Belongie, S.: Robust object tracking with online multiple instance learning. IEEE Trans. Pattern Anal. Mach. Intell. 33(8), 1619–1632 (2011)

138. Cetin, A.E., Akhan, M.B., Toreyin, B.U.: Characterization of Motion of Moving Objects in Video. United States Patent Application 20040223652, (2004)
139. Toreyin, B.U., Cetin, A.E., Aksay, A., Akhan, M.B.: Moving region detection in compressed video. IEEE Int. Conf. Comput. Inf. Sci. 3280, 381–390 (2004)
140. Toreyin, B.U., Cetin, A.E., Aksay, A., Akhan, M.B.: Moving object detection in wavelet compressed video. J. Signal Process. Image Commun. 20(3), 255–264 (2005)
141. Cheng, F.H., Chen, Y.L.: Real time multiple objects tracking and identification based on discrete wavelet transform. J. Pattern Recognit. 39(6), 1126–1139 (2006)
142. Schutte, H.D., Wenzel, J.: Hypercomplex numbers in digital signal processing. IEEE Int. Conf. Circuits Syst. 2, 1557–1560 (1990)

Further Quaternion Integral Transforms

Eckhard Hitzer

Chapter Introduction As may be expected the modification of the QFT to a localized QWT only formed the beginning of a rich field of novel quaternionic integral transforms, that often are motivated by generalizing known scalar integral transforms to higher dimensions, taking advantage of how quaternions geometrically meaningfully and efficiently combine several signal components. This chapter is devoted to survey these new types of quaternionic integral transforms.

Wavelets of particular shape are known as curvelets, ridgelets and shearlets. Their generalization to quaternionic versions is introduced in Sect. 1.

A major generalization of Fourier transforms are linear canonical transforms subsuming Fourier transforms, fractional Fourier transforms, (fractional) Laplace transforms, scaling, Fresnel transforms, chirp multiplication, Gauss-Weierstrass transforms and Bargmann transforms, etc. They have wide applications in fields like optics and quantum mechanics. By now a lot of work has been done to study *quaternionic linear canonical transforms* (QLCT), which we survey in Sect. 2.

Section 2.1 features QLCT related theorems. Section 2.2 focuses on the study of (windowed) QLCT and uncertainty. Section 2.3 shows how the QLCT is applied to sampling of quaternionic signals and to envelop detection of generalized analytic signals. Section 2.4 covers further mathematical developments of QLCT related transforms, like offset QLCT (special affine quaternion Fourier transforms), Wigner-Ville distributions, wave packet transforms and short-time transforms.

Section 3 surveys other *quaternionic integral transforms* (QIT) beyond QWT and QLCT. Section 3.1 covers quite a variety of new theoretical developments of interest in special fields of hypercomplex algebra and hypercomplex analysis. Section 3.2 shows how other QITs are applied, including vector field analysis, and prolate spheroidal wave function sampling.

E. Hitzer (✉)
International Christian University, Mitaka, Tokyo, Japan
e-mail: hitzer@icu.ac.jp

E. Hitzer (ed.), *Quaternionic Integral Transforms*, Trends in Mathematics,
https://doi.org/10.1007/978-3-031-28375-8_3

Machine Generated Keywords Canonical, linear canonical, canonical transform, QLCT, uncertainty, uncertainty principle, principle, linear, formula, Heisenberg, quaternion linear, two-side, integral transform, LCT, group

1 Quaternionic Curvelets, Ridgelets and Shearlets

Machine Generated Keywords Ridgelet, shearlet transform, shearlet, uncertainty principle, uncertainty, principle, reconstruction formula, transform associate, associate quaternion, target, reconstruction, transform, element, kernel, article

1.1 Uncertainty Principles for the Continuous Quaternion Shearlet Transform

DOI: https://doi.org/10.1007/s00006-019-0961-4

Abstract-Summary
We present some new elements of harmonic analysis related to the continuous quaternion shearlet transform. The main objective of this article is to introduce the concept of the quaternionic Shearlet transform and investigate its different properties using the machinery of quaternion Fourier transforms and quaternion convolution.

Introduction
We present some new elements of harmonic analysis related to the continuous quaternion shearlet transform and we establish uncertainty principles for this transform.

We analyse the concentration of the quaternion shearlet transform on sets of finite measure using appropriate projection operators.

Donoho-Stark and local-type uncertainty principles are given.

We give a general form of the Heisenberg type inequality for the continuous quaternion shearlet transform.

Generalities
We recall some basic definitions and properties of the quaternion Fourier transform.

The quaternion algebra was formally introduced by the Irish mathematician W.R Hamilton in 1843, and it is a generalization of complex numbers.

In [1], the authors gave a demonstration of the property by considering some other assumptions.

Continuous Quaternion Shearlet Transform
The shearlet transform was originally developed in the inaugural paper [2] and became a very useful mathematical tool, which has been widely used in

characterization of function spaces as well as in signal and image processing, see [3, 4, 5, 6–7, 8, 9, 10].

For more details on the shearlet transform, the reader can see [4, 11].

[Section 4]

The classical uncertainty principle was established by Heisenberg in [12] raising a fundamental problem in quantum mechanics: Both the position and the momentum of particles cannot be determined explicitly but only in a probabilistic sense with a certain 'uncertainty'.

Heisenberg did not give a precise mathematical formulation of the uncertainty principle, but this was done in the late 1920s by Kennard [13] and Weyl (who attributes the result to Pauli) [[14], Appendix 1].

Acknowledgement

A machine generated summary based on the work of Brahim, Kamel; Nefzi, Bochra; Tefjeni, Emna

2019 in Advances in Applied Clifford Algebras

1.2 Quaternion Ridgelet Transform and Curvelet Transform

DOI: https://doi.org/10.1007/s00006-018-0897-0

Original Abstract In this article, we study the quaternion ridgelet transform and curvelet transform associated to the quaternion Fourier transform (QFT). We prove some properties related to such transforms, including reconstruction formulas, reproducing kernels and uncertainty principles.

Acknowledgement

A machine generated summary based on the work of Ma, Guangsheng; Zhao, Jiman

2018 in Advances in Applied Clifford Algebras

1.3 Infrared Small Target Detection Based on Non-subsampled Shearlet Transform and Phase Spectrum of Quaternion Fourier Transform

DOI: https://doi.org/10.1007/s11082-020-02292-x

Abstract-Summary

Inspired by previous studies showing that the phase spectrum of the quaternion Fourier transform (PQFT) has great superiority in salient region extraction and the desirable characteristics of multi-scale, multi-direction and shift-invariance of the non-subsampled shearlet transform (NSST), a new target detection method is proposed based on NSST and PQFT in this paper.

The original image is first subjected to NSST decomposition to obtain a low frequency component and four high frequency components by NSP.

Directional localization is achieved by shearing filters which provides multi-directional decomposition.

Four direction high frequency sub-images decomposed by NSST are introduced as four data channels of the PQFT.

The real target is directly segmented by an adaptive threshold.

Introduction

The targets are apt to immerse in background clutter which results in some detection algorithms being not ideal for detecting targets with low signal-to-noise ratio (SNR).

Because time domain information easily causes false alarm and is time-consuming in infrared image sequences, many researchers concentrate on turning infrared small target detection to single-frame detection.

For the filtering methods, fixed templates are used to highlight a target and to suppress the background directly.

Another far-reaching saliency detection method is based on the Fourier transform (Guo and others 15), it can extract the remarkable target area through phase spectrum information and achieve a preferable effect.

A new infrared small target detection method combined with NSST and PQFT is presented in this paper, in which the NSST is introduced to build four data channels.

Since the background in the image is mainly distributed in the low frequency component part, the target always appears in the high frequency components.

Relevant Theories

The key to target detection is to effectively distinguish the background edge and other regions with similar gradient step characteristics from the small target, and effectively remove the small target signal while retaining the background to obtain an accurate background image.

Details such as weak targets and background edges are concentrated in the high frequency sub-band.

There is still some clutter in the decomposed image, and further fusion of all high frequency sub-bands is needed to achieve the purpose of enhancing the target and suppressing background clutter.

This paper combines the PQFT algorithm to process the high frequency sub-bands, and obtains the sub-bands of the weakened small target signals after suppression.

The image is segmented by a threshold method to separate the target after obtaining the reconstructed map TEM.

Experimental Results and Analysis

Although all five algorithms detect the target, BHF, MAX-median, and TDLMS simultaneously detect a large amount of clutter.

Top-hat suppresses the background clutter to a certain extent and highlights the target.

The five were submerged by clutter, and Max-median, TDLMS, Top-hat did not detect the target.

All targets are successfully detected in all five test sequences and the clutter suppression effect is the best.

Therefore, our method based on a non-subsampled shearlet transform really has better performance in infrared small target detection.

LSBRG is an improvement of LSBR, which can be utilized to evaluate the capability of target/background contrast enhancement under different signal-to-noise ratios (SNRs).

The higher the value of LSBRG, the better the performance of our algorithm detection target.

Our method possesses more effectiveness and robustness under different backgrounds and target types compared with the other four conventional methods.

Conclusion

A new infrared small target detection method based on NSST and PQFT is presented.

Several experiments test the detection accuracy and computational efficiency in the complex background of typical scenes such as sky–cloud, ground and sea–sky.

Through qualitative evaluation and quantitative evaluation, the LSBRG is the highest and the ROC area is the largest compared with the other four classical detection algorithms.

It is proved that our method has the highest detection accuracy and the best detection effect.

The algorithm running speed can also basically reach the real-time requirement.

Acknowledgement

A machine generated summary based on the work of Ren, Kan; Song, Congcong; Miao, Xin; Wan, Minjie; Xiao, Junfeng; Gu, Guohua; Chen, Qian

2020 in Optical and Quantum Electronics

2 Quaternionic Linear Canonical Transforms (QLCT)

Machine Generated Keywords Linear canonical, canonical, canonical transform, QLCT, linear, quaternion linear, LCT, QLCT domain, uncertainty, uncertainty principle, principle, two-side QLCT, two-side, offset, offset linear

2.1 QLCT Theorems

Machine Generated Keywords Canonical transform, linear canonical, canonical, LCT, QLCT, quaternion linear, linear, Hilbert transform, edge, Hilbert, edge detection, transform quaternionic, integral transform, direct, approximation

2.1.1 Plancherel Theorems of Quaternion Hilbert Transforms Associated with Linear Canonical Transforms

DOI: https://doi.org/10.1007/s00006-019-1034-4

Abstract-Summary
The quaternion Hilbert transforms associated with the linear canonical transform are recently used to form the quaternion analytic signal.

Some properties of the 2D quaternion Hilbert transforms with the two-sided quaternion linear canonical transforms are investigated, such as the Plancherel theorems, the Parseval identities and the inversion formulas of the Hilbert transforms.

We define the discrete generalized quaternion Hilbert transforms and use them for color edge detection.

Introduction
With the development of the linear canonical transform (LCT), the 1D and 2D HTs have been extended into LCT domains [16, 17, 18].

The theory on the 2D analytic signal in the quaternion linear canonical transform (QLCT) domain was studied in [19].

We study the properties of the 2D HT associated with the two-sided QLCT.

We also introduce the definitions of the linear canonical transforms and the quaternion linear canonical transforms (QLCT).

Preliminaries
The present section collects some basic facts about quaternions, QFTs, QLCTS, the definitions of Hilbert transforms (HT) of 2D real-valued functions and the definitions of the 2D quaternion Hilbert transforms (QHT) in the QLCT domain, which will be needed throughout the text.

Due to the non-commutativity of the quaternionic algebra, there are different types of the QFT.

The Plancherel theorem and the uncertainty principles for the right-sided QLCT of 2D quaternionic signals are derived in [20].

Applications to 2D Edge Detection
We then define the discrete version of the generalized QHT (GQHT), which will be used to detect the edges of color images in the subsequent section (DGQHT). The discrete GQHTs are defined where DLCT means the discrete LCT and IDLCT means the inverse DLCT.

We apply the proposed DGQHT to edge detection in color images.

The key difference between the work [21] and ours is that the work [21] is confined to the edge detection of grayscale images while the proposed method in this work is designed for color image edge detection by regarding the color image in a holistic manner.

Differentiation detects the edge of an image by detecting the difference between the neighboring pixels.

The results demonstrate the capability of the DGQHT for edge detection in color images.

It shows that our proposed method can successfully detect most of the edges in this image.

Conclusion

Several properties of the 1D Hilbert transform associated Fourier transform are extended to the 2D Hilbert transforms and the quaternion Hilbert transform by using the two-sided quaternion Fourier transform and the quaternion linear canonical transform, such as the Plancherel theorems, the Parseval identities and the inversion formulas of the Hilbert transforms.

Several properties of the quaternion Hilbert transforms (QHTs) are provided.

Acknowledgement

A machine generated summary based on the work of Kou, Kit Ian; Liu, Ming-Sheng; Zou, Cuiming

2019 in Advances in Applied Clifford Algebras.

2.1.2 Jackson Theorems for the Quaternion Linear Canonical Transform

DOI: https://doi.org/10.1007/s00006-022-01226-y

Introduction

The linear canonical transform is an integral transform which serves as a generalization of the Fourier transform, the fractional Fourier transform and the Fresnel transform [22, 23], was firstly proposed by Moshinsky and Collins [24, 25].

The quaternion algebra offers a simple and profound representation of signals wherein several components are controlled simultaneously.

The development of integral transforms for quaternion-valued signals has found numerous applications in science and engineering [26].

Until now, many integral transforms have been extended in the realm of quaternion algebra, including quaternion Fourier and wavelet transforms [27], the quaternionic ridgelet transform [28], the quaternionic Stockwell transform [29], the quaternionic curvelet transform [30], the quaternionic linear canonical transforms [31] and many more [32, 33–34].

Motivated and inspired by the above work, we in this paper establish Bernstein's inequality, and Jackson's direct and inverse theorems for the quaternion linear canonical transform.

Preliminaries
We recall the basics of quaternion algebra, the quaternion linear canonical transform along with its basic properties and some auxillary results that will be used to prove the main results in this paper.

The fundamental properties of the QLCT can be found in [35].

Bernstein's Inequality and Jackson's Direct Theorems
We will state and prove Bernstein's inequality, and Jackson direct and inverse theorems for the QLCT.

Acknowledgement
A machine generated summary based on the work of Achak, A.; Ahmad, O.; Belkhadir, A.; Daher, R.

2022 in Advances in Applied Clifford Algebras

2.2 QLCT and Uncertainty

Machine Generated Keywords QLCT, uncertainty, uncertainty principle, canonical transform, linear canonical, principle, canonical, twoside QLCT, twoside, quaternion linear, QLCT domain, principle twoside, twosided, quaternion-valued signal, logarithmic uncertainty

2.2.1 Novel Uncertainty Principles for Two-Sided Quaternion Linear Canonical Transform

DOI: https://doi.org/10.1007/s00006-018-0828-0

Abstract-Summary
The uncertainty principle, which offers information about a function and its Fourier transform in the time-frequency plane, is particularly powerful in mathematics, physics and for the signal processing community.

Based on the fundamental relationship between the quaternion linear canonical transform (QLCT) and the quaternion Fourier transform (QFT), we propose two different uncertainty principles for the two-sided QLCT.

It is shown that lower bounds can be obtained on the product of spreads of a quaternion-valued function and its two-sided QLCT from newly derived results.

Acknowledgement
A machine generated summary based on the work of Zhang, Yan-Na; Li, Bing-Zhao

2018 in Advances in Applied Clifford Algebras

2.2.2 On Uncertainty Principle for the Two-Sided Quaternion Linear Canonical Transform

DOI: https://doi.org/10.1007/s11868-021-00395-x

Abstract-Summary
The quaternion linear canonical transform (QLCT), as a generalized form of the quaternion Fourier transform, is a powerful analyzing tool in image and signal processing.

Introduction
The uncertainty principle, which provides a lower bound on the spreads of two specific transform domains, is of importance in various scientific fields such as mathematics [36, 37], physics [38, 39] and signal processing [40, 41, 42].

In harmonic analysis, the uncertainty principle states that a non-zero function and its Fourier transform cannot both be very rapidly decreasing.

Zhang and Li [43] described lower bounds of spreads of a non-zero quaternion-valued signal function in two different QLCT domains, which include those uncertainty principles of one quaternion-valued signal in the spatial and frequency domain as its special case.

We are not only to show the uncertainty principles of one quaternion-valued signal in spatial and frequency domains, but also to extend the uncertainty principles to one quaternion-valued signal in the spatial and another quaternion-valued signal in the frequency domain.

Novel derived uncertainty principles in the QLCT domains are expected to further contribute to solving the problem of quaternion-valued signal recovery in practical applications.

Preliminaries
We recall some basic notation and related facts, which are used in the quaternion algebra, the two-sided quaternion Fourier transform (QFT) and the two-sided quaternion linear canonical transform (QLCT).

There are usually three different types of QFT, the left-sided QFT, the two-sided QFT, and the right-sided QFT (cf. [44]).

This is why we focus on the two-sided QLCT instead of the left-sided (or right-sided) QLCT.

Uncertainty Principle
We mainly use some uncertainty principles for the two-sided QFT to prove the corresponding uncertainty principles for the two-sided QLCT.

In [45], the authors proved the logarithmic uncertainty principle for a real valued function under the two-sided QFT following Beckner's line with quite a lot of efforts.

Lian [46] in 2018 presented the logarithmic uncertainty principle for a quaternion valued function under the two-sided QFT based on the orthogonal two-dimensional plane split of quaternions.

The purpose of this subsection is to establish the logarithmic uncertainty principle for a quaternion valued function under the two-sided QLCT.

We first recall the Heisenberg uncertainty principle for the two-sided QFT.

The entropic uncertainty principle is a fundamental tool in information theory, quantum physics and harmonic analysis.

We next propose the entropic uncertainty principle for the two-sided QLCT.

Conclusions

Using the relationship between the two-sided QLCT and QFT, we obtain five different forms of uncertainty principles associated with the two-sided QLCT.

The uncertainty principles for the two-sided QFT and FrQFT are special cases.

We expect that our generalization of the uncertainty principles for the two-sided QLCT will be of great significance in practical application.

Acknowledgement

A machine generated summary based on the work of Zhu, Xiaoyu; Zheng, Shenzhou 2021 in Journal of Pseudo-Differential Operators and Applications

2.2.3 Uncertainty Principles for the Two-Sided Quaternion Linear Canonical Transform

DOI: https://doi.org/10.1007/s00034-020-01376-z

Abstract-Summary

The quaternion linear canonical transform (QLCT), as a generalized form of the quaternion Fourier transform, is a powerful analyzing tool in image and signal processing.

In order to analyze a non-stationary signal and a time-varying system, we present Lieb's uncertainty principle for the two-sided short-time quaternion linear canonical transform (SQLCT) based on the Hausdorff–Young inequality.

By adding the nonzero quaternion-valued window function, the two-sided SQLCT has a very significant application in the study of the local signal local frequency spectrum.

Introduction

The uncertainty principle first introduced by German physicist Heisenberg in [12] plays an important role in signal processing, physics and mathematics.

They only considered the uncertainty principles on a nonzero quaternion-valued function and its QLCT domain.

We would like to mention that Zhang and Li [47] extended the Heisenberg uncertainty principle to a nonzero quaternion-valued signal function in two different QLCT domains so that it led to a big breakthrough in signal recovery.

We not only show the uncertainty principles of one quaternion-valued signal in the spatial and frequency domains, but also extend the uncertainty principles to one quaternion-valued the signal in the spatial and another quaternion-valued signal in the frequency domain.

In order to analyze a non-stationary signal and a time-varying system, we also present Lieb's uncertainty principle for the two-sided SQLCT.

We prove Lieb's uncertainty principle for the two-sided short-time quaternion linear canonical transform (SQLCT).

Preliminaries

We recall basic notations and some related facts, which are involved in quaternion algebra, the two-sided QFT and the two-sided QLCT.

There are usually three different types of QFT: the left-sided QFT, the two-sided QFT and the right-sided QFT (cf. [44]).

This is why we focus on the two-sided QLCT instead of the left-sided (or right-sided) QLCT.

Uncertainty Principles

The aim of this subsection is to extend Hardy's uncertainty principle to that for the two-sided QLCT.

Before stating our first result, we recall Hardy's uncertainty principle for the two-sided QFT which is illustrated in [48].

It can be seen that Hardy's uncertainty principle for the two-sided QLCT is the generalization of Hardy's uncertainty principle for the two-sided QFT.

Of all, let us recall Beurling's uncertainty principle for the two-sided QFT which is interpreted in [49].

We extend it to that for the quaternion region, which means that we will establish the uncertainty principle for the two-sided short-time quaternion linear canonical transform (SQLCT).

We now present Lieb's uncertainty principle for the two-sided SQLCT.

Potential Applications

As for a generalization of a transformation from complex to quaternion algebra, the QLCT transforms a quaternionic 2D signal into a quaternion-valued frequency domain signal, which is an effective processing tool for color image analysis.

The new uncertainty principles for the two-sided QLCT describe the relation of one quaternion-valued signal in the spatial and another quaternion-valued signal in the frequency domain.

They can further contribute to solving problems of color image processing, quantum mechanics, electromagnetism, signal processing, optics, electrodynamics, etc. Specially, Donoho–Stark's uncertainty principle for the two-sided QLCT has great applications in signal reconstruction from a noisy measurement.

Conclusions

We obtain three different forms of uncertainty principles associated with the two-sided QLCT, and finally we investigate Lieb's uncertainty principle for the two-sided SQLCT.

The uncertainty principle for the two-sided QFT or FRQFT is as a special of that for the two-sided QLCT.

We expect that our generalization of the uncertainty principles for the two-sided QLCT will be of great significance in practical application.

Acknowledgement
A machine generated summary based on the work of Zhu, Xiaoyu; Zheng, Shenzhou 2020 in Circuits, Systems, and Signal Processing

2.2.4 Uncertainty Principle for the Two-Sided Quaternion Windowed Linear Canonical Transform

DOI: https://doi.org/10.1007/s00034-021-01841-3

Abstract-Summary
We investigate the (two-sided) quaternion windowed linear canonical transform (QWLCT) and study the uncertainty principles associated with the QWLCT.

We study different kinds of uncertainty principles for the QWLCT, such as the logarithmic uncertainty principle, the entropic uncertainty principle, the Lieb uncertainty principle and Donoho–Stark's uncertainty principle.

We provide a numerical example and a potential application to signal recovery by using Donoho–Stark's uncertainty principle associated with the QWLCT.

Extended:

We will discuss the physical significance and the engineering background of this paper in the future.

Introduction
The QLCT was first studied in [50] including prolate spheroidal wave signals and uncertainty principles.

Some useful properties of the QLCT, such as linearity, the reconstruction formula, continuity, boundedness, the positivity inversion formula and the uncertainty principle, were established in [51, 52, 53, 54–55, 56].

In signal processing, the uncertainty principle states that the product of the variances of a signal in the time and frequency domains has a lower bound.

The Lieb uncertainty principle for the WLCT was discussed in [57], which has the LCT version as one of its special cases.

Huang and others [58] discussed the uncertainty principle and the orthogonal condition for the WLCT in one WLCT domain.

In [59], based on the real window function, Heisenberg's uncertainty principle for the QWLCT was obtained.

The aim of this paper is to obtain several uncertainty principles for the QWLCT.

Preliminaries
We mainly review some basic facts of quaternion algebra and the QLCT, which will be needed throughout the paper.

The QLCT was first defined by Kou and others, and it is a generalization of the LCT in the framework of quaternion algebra [60, 61].

We mainly focus on the two-sided QLCT.

Quaternionic Windowed Linear Canonical Transform (QWLCT)
The generalization of the window function associated with the QLCT is discussed, which is denoted as QWLCT.

This subsection leads to a 2D quaternion window function associated with the QLCT.

Due to the noncommutative property of the multiplication of quaternions, there are three different types of QWLCTs: the two-sided QWLCT, left-sided QWLCT and right-sided QWLCT.

Uncertainty Principles for the QWLCT
Based on Pitt's inequality, the logarithmic uncertainty principle for the QFT was proved in [62].

We present the entropic uncertainty principle for the QWLCT.

The entropic uncertainty principle is very important in signal processing and harmonic analysis.

Numerical Example and Potential Application
The uncertainty principles for the QWLCT have been studied in this paper.

The uncertainty principle has some applications in signal recovery.

The authors of [63] have extended signal recovery by using a local uncertainty principle for the QLCT.

We will give a potential application for signal recovery by using Donoho–Stark's uncertainty principle for the QWLCT.

Conclusions
Pitt's inequality and the Lieb inequality for the QWLCT are investigated and different forms of uncertainty principles associated with the QWLCT are proposed.

Based on the relationship between the QWLCT and the QFT, Pitt's inequality and the Lieb inequality associated with the QWLCT are demonstrated.

The uncertainty principles for the QWLCT, such as the logarithmic uncertainty principle, the entropic uncertainty principle, the Lieb uncertainty principle and Donoho–Stark's uncertainty principle are obtained.

Further investigations on this topic are now underway, such as on the directional uncertainty principle for the QWLCT.

Acknowledgement
A machine generated summary based on the work of Gao, Wen-Biao; Li, Bing-Zhao 2021 in Circuits, Systems, and Signal Processing

2.3 *QLCT Applications*

Machine Generated Keywords QLCT, canonical transform, linear canonical, canonical, signal, sample, quaternionic signal, sample formula, analytic signal, analytic, signal paper, Hilbert transform, Hilbert, LCT, theorem

2.3.1 Sampling Formulas for Non-bandlimited Quaternionic Signals

DOI: https://doi.org/10.1007/s11760-021-02110-1

Abstract-Summary
Sampling theorems for certain types of non-bandlimited quaternionic signals are proposed.

Introduction
Enlightened by [64, 65], the authors in [66] structured a two-parameter ladder-shaped filter in the linear canonical transform domain and established sampling theorems for NBL real signals in the linear canonical transform sense.

Using quaternion prolate spheroidal wave functions, sampling theorems for 2D BL quaternionic signals were first derived in [67].

Sampling theorems for 2D BL quaternionic signals associated with right-sided QFT and QLCT were proposed in [68].

In [69], the authors have derived sampling theorems of 2D BL quaternionic signals under various QFTs and QLCTs.

By the generalized sinc functions, we will first establish the spaces of NBL quaternionic signals, and then the sampling theorems for two classes of NBL quaternionic signals are established.

We propose four types of sampling formulas for the NBL quaternionic signals associated with the QLCT and QFT, respectively.

Preliminaries
This section is devoted to the exposition of basic preliminary material which we use extensively throughout of this paper.

Based on the concept of quaternion, many types of QFTs and QLCTs [70] have been introduced.

Sampling Theorems for NBL Quaternionic Signals
We deal with quaternionic signals which are non-bandlimited in the frequency domain.

We introduce a function space consisting of NBL quaternionic signals with special spectrum properties associated with a ladder-shaped filter of two parameters.

Experimental Results
We also compare our methods with the sampling formula for quaternionic signal in [69].

It is seen that the proposed sampling formulas can be applied effectively in the reconstruction of non-bandlimited signals.

It can be seen that the sampling formulas for non-bandlimited signals outperform the sampling formula for bandlimited signals [69].

For quaternionic signal f(x), 250 sampling points are needed for reconstruction, while 150 sampling points are enough for the reconstruction of the quaternionic signal g(x).

Conclusion and Discussion

These sampling formulas are applicable for certain types of the non-bandlimited quaternionic signals.

According to the quaternion Riemann-Lebesgue lemma [70], a majority of non-bandlimited quaternionic signals satisfy the requirement of the proposed theorems approximately.

Quaternions are gaining popularity in signal and image processing (see e.g., [71, 72]).

There is a need to develop the 2D sampling series of non-bandlimited quaternionic signals to provide an effective tool in image processing.

Acknowledgement

A machine generated summary based on the work of Hu, Xiaoxiao; Kou, Kit Ian 2022 in Signal, Image and Video Processing

2.3.2 Envelope Detection Using Generalized Analytic Signal in 2D QLCT Domains

DOI: https://doi.org/10.1007/s11045-016-0410-7

Abstract-Summary

The hypercomplex 2D analytic signal has been proposed by several authors with applications in color image processing.

The analytic signal enables to extract local features from images.

The extension of the analytic signal of the linear canonical transform domain from 1D to 2D, covering also intrinsic 2D structures, has been proposed.

The quaternion Fourier transform plays a vital role in the representation of multidimensional signals.

The quaternion linear canonical transform (QLCT) is a well-known generalization of the quaternion Fourier transform.

Different approaches to the 2D quaternion Hilbert transforms are proposed that allow the calculation of the associated analytic signals, which can suppress the negative frequency components in the QLCT domain.

The analytic signal is a complex signal derived by adding the real signal in quadrature with its Hilbert transform.

Introduction

A first definition of a 2D quaternion linear canonical transform (QLCT) was introduced in Kou et al. 73 and Kou et al. 20, which is discussed later in this paper, and the first application of a 2D quaternion linear canonical transform to multidimensional signal analysis has been reported in Kou et al. 73, Kou et al. 20 and Kou et al. 55 involving prolate spheroidal wave signals and uncertainty principles (Yang and Kou 74).

The main goals of the present paper are to study the properties of the two-sided QLCTs of 2D quaternionic signals and to derive the novel concept of quaternion Hilbert transform (QHT).

We study the two-sided QLCT of 2D quaternionic signals.

In quaternionic analysis, the quaternion Fourier transform (QFT) (Chen et al. 45; Hitzer 61; Mawardi et al. 75, 76; Pei et al. 44) plays a vital role in the representation of (multidimensional) signals.

It transforms a real (or quaternionic) 2D signal into a quaternion-valued frequency domain signal.

In Kou and Morais 52, the authors established a generalized Riemann–Lebesgue lemma for the (right-sided) QLCT, which prescribes the asymptotic behaviour of the QLCT extending and refining the classical Riemann–Lebesgue lemma for the Fourier transform of 2D quaternion signals.

Preliminaries

The present section collects some basic facts about quaternions and the two-sided QFT, which will be needed throughout the text.

We recall the QFT, which will be needed to establish the QLCT in the next section.

Due to the non-commutativity of the underlying quaternion multiplication, there is no left- or right-linearity property for general linear combinations with quaternionic constants.

There is a left-linearity property for general linear combinations with complex constants; some other fundamental properties of the two-sided QFT were given (Hitzer 61) in detail.

[Section 3]

The above two definitions have been applied in quaternion wavelet analysis with applications to image denoising (Yin et al. 77).

The two-sided QFT allows us to pass from the space domain to the frequency domain.

The requirement that the original signal be reconstructible from the analytic signal is fulfilled.

The Quaternion Linear Canonical Transform

Due to the non-commutativity of the multiplication of quaternions, the definition of QLCT is not unique and several approaches of introducing it have been developed (Kou et al. 20, 55).

The right-sided QLCT of 2D quaternionic signals (Kou et al. 20) was studied in detail.

We study the two-sided QLCTs of 2D quaternionic signals (Kou et al. 55).

The QHT and GQAS

The study of multidimensional Hilbert transforms in the QLCT domain has not been carried out yet.

One reason is that 2D quaternionic signals have more complex structures than 2D real signals, and another reason is that there exist several different 2D Hilbert transforms in the LCT domain.

Researches are developing multidimensional Hilbert transforms in the QFT domain (Bülow and Sommer 78; Ell and Sangwine 79).

In the present section, the 2D left-sided, right-sided and two-sided Hilbert transforms in the QLCT domain are presented.

Examples on the Envelope Detection
The envelope of a quaternionic signal is obtained as the magnitude of its associated GQAS, and this envelope is the instantaneous amplitude of the GQAS.

Envelope detection for an approach towards 2D complex or quaternionic signals are based on different combinations of the original signal and its partial and total Hilbert transforms.

The reader can clearly see that the envelope was successfully extracted from the sample signal.

The GQAS is a quaternionic-valued signal, where the scalar part is the original signal and the non-scalar part is the QHT of the original signal.

The reader can clearly see that the envelope was successfully extracted from the sample signal chosen from different quadrants.

Outlook and Concluding Remarks
The two-sided quaternion linear canonical transform (QLCT) of 2D quaternionic signals for envelope detection is proposed.

Some elementary properties and the inversion of the QLCT are discussed.

Due to the non-commutative nature of quaternions, we introduce the concept of QHT by means of the left-sided, right-sided and two-sided of QLCTs.

Acknowledgement
A machine generated summary based on the work of Kou, Kit Ian; Liu, Ming-Sheng; Morais, João Pedro; Zou, Cuiming

2016 in Multidimensional Systems and Signal Processing

2.4 *QLCT Related Transforms*

Machine Generated Keywords linear canonical, canonical, offset, canonical transform, offset linear, linear, distribution, quaternion offset, integral transform, associate quaternion, QLCT, nonstationary, Bahri, wave, integral

2.4.1 Short-Time Special Affine Fourier Transform for Quaternion-Valued Functions

DOI: https://doi.org/10.1007/s13398-022-01210-y

Abstract-Summary

The special affine Fourier transform is a promising tool for analyzing transient signals with more degrees of freedom via a chirp-like basis.

Our goal is to introduce a novel quaternion-valued short-time special affine Fourier transform in the context of two-dimensional quaternion-valued signals.

Introduction

Despite humongous merits of the quaternion special affine Fourier transform, it fails in providing an adequate time-frequency representation of non-stationary signals as the associated kernel is global in nature.

Our goal is to evade such limitations of the quaternion special affine Fourier transform by formulating a novel integral transform coined as the quaternion-valued short-time special affine Fourier transform, which relies upon a sliding window to capture the localized spectral contents of non-stationary two-dimensional quaternion-valued signals.

It is imperative to mention that due to the non-commutativity of quaternions, several different versions of the quaternion special affine Fourier transforms are available in the literature [80, 81], however, we are propelled by the approach due to Haoui and others [80].

Our goal is to include certain reasonable localization features to the said quaternion special affine Fourier transform by windowing the spectral contents via a two-dimensional window function.

Preliminaries

The quaternion algebra is an associative, non-commutative four-dimensional algebra that extends the complex number system.

The quaternion version of Cauchy-Schwartz's inequality is given. Due to the non-commutativity of quaternion multiplication, there are three types of the quaternion special affine Fourier transform, the left-sided QSAFT, the right-sided QSAFT, and two-sided QSAFT.

Uncertainty Principles for the Quaternion-Valued Short-Time SAFT

The classical Heisenberg uncertainty principles in harmonic analysis state that a nonzero signal and its Fourier transform cannot be both localized simultaneously [82].

There are many expressions of this general fact where the localization and the smallness are interpreted differently and in several ways, for instance, a logarithmic version of the uncertainty principle obtained by Beckner [83] while using a sharp form of Pitt's inequality.

The local version of uncertainty states that if a function is concentrated, then not only is its transformation spread out, but that it cannot be localized in a subset of finite measure (see [46, 84, 85]).

Concluding Remarks and Observations

We have introduced a new time-frequency tool for analyzing two-dimensional quaternion-valued signals namely, the quaternion-valued short-time special affine Fourier transform.

Some important properties such as inner product relation, energy conservation and inversion formula are derived for the quaternion-valued short-time special affine Fourier transform.

Several uncertainty principles, including Heisenberg-Weyl, logarithmic and local-type uncertainty principles, have also been derived for the quaternion-valued short-time special affine Fourier transform.

In concluding our present investigation, we choose to draw the reader's attention toward some recent developments (see [86, 87] and [88]) on the uncertainty principles, and the wavelet and other transforms.

Acknowledgement
A machine generated summary based on the work of Srivastava, H. M.; Shah, Firdous A.; Teali, Aajaz A.

2022 in Revista de la Real Academia de Ciencias Exactas, Físicas y Naturales. Serie A. Matemáticas

2.4.2 Convolution and Correlation Theorems for Wigner–Ville Distribution Associated with the Quaternion Offset Linear Canonical Transform

DOI: https://doi.org/10.1007/s11760-021-02074-2

Abstract-Summary
The quaternion offset linear canonical transform (QOLCT) has gained much popularity in recent years because of its applications in many areas, including image and signal processing.

The applications of the Wigner–Ville distribution (WVD) in signal analysis and image processing cannot be excluded.

Introduction
In time–frequency signal analysis, the classical Wigner–Ville distribution (WVD) or Wigner–Ville transform (WVT) has an important role to play.

The convolution has numerous important applications in various areas of mathematics like linear algebra, numerical analysis and signal processing, whereas correlation like convolution is an another important tool in signal processing, optics and detection applications.

Motivated by the generalized LCT, the quaternion offset linear canonical transform and the Wigner–Ville distribution, in this paper we study the Wigner–Ville distribution associated with quaternion offset linear canonical transform (WVD-QOLCT).

A novel canonical convolution operator and a related correlation operator for WVD-QOLCT are proposed.

Wigner–Ville Distribution Associated with Quaternion Offset Linear Canonical Transform (WVD-QOLCT)
Further some basic properties of the WVD-QOLCT, which are important for signal representation in signal processing, are investigated.

These properties play important roles in signal representation.

The following theorem guarantees the reconstruction of the input quaternion signal from the corresponding WVD-QOLCT within a constant factor.

Convolution and Correlation Theorem for WVD-QOLCT
We first define the convolution and correlation for the QOLCT.

We then establish the new convolution and correlation for the WVD-QOLCT.

The convolution and correlation theorems will open new gates to investigate the sampling and filtering theorems of the WVD-QOLCT.

As a consequence of the above definition, we get the following important theorem, which states how the convolution of two quaternion-valued functions interacts with their QOLCTs.

Applications
Since the WVD-QOLCT is a generalization of the WVD-QLCT and the QWVD, it is customary to study its application to the detection of a quaternion-valued LFM signal.

We present an example to show how the WVD-QOLCT is applied to form the LFM of quaternion signals.

It is clear from the above illustration that we can detect non-stationery LMF quaternion-valued signals with the WVD-QOLCT.

Acknowledgement
A machine generated summary based on the work of Bhat, M. Younus; Dar, Aamir H. 2022 in Signal, Image and Video Processing

2.4.3 Uncertainty Principles for Wigner–Ville Distribution Associated with the Quaternion Offset Linear Canonical Transform

DOI: https://doi.org/10.1007/s00034-022-02127-y

Abstract-Summary
The Wigner–Ville distribution (WVD) associated with the quaternion offset linear canonical transform (QOLCT) (WVD–QOLCT) is the known furthest generalization of the WVD in quaternion algebra.

Some properties and the classical Heisenberg uncertainty principle (UP) have been derived for the two-dimensional (2D) two-sided WVD–QOLCT.

By using the nonlinearity property, applications of the 2D WVD–QOLCT in linear frequency modulated signal detection are proposed.

Introduction

In [80], a new quaternion integral transform known as the QOLCT has been introduced by El Haoui and Hitzer, which generalizes the offset linear canonical transform (OLCT) for the quaternion algebra.

The QOLCT transforms a quaternionic 2D signal into a quaternion-valued frequency-domain signal.

The Wigner–Ville distribution associated with the quaternion offset linear canonical transform (WVD–QOLCT) is therefore proposed in [89, 90] to overcome this drawback and to be a flexible transform that combines both the results and the flexibility of the two transforms, Wigner–Ville distribution (WVD) and QOLCT.

The authors in [89, 90] derived some important properties of this transform and established a version of Heisenberg's inequality, Lieb's theorem, and the Poisson summation formula for the WVD–QOLCT.

The WVD–QOLCT may be seen as a continuation of a wide number of papers that have been devoted to the extension of the theory of the windowed FT to the quaternionic case: Hahn and Snopek constructed a Fourier-Wigner distribution of 2D quaternionic signals [91].

Wigner–Ville Distribution in the Quaternion Offset Linear Canonical Transform Domain (WVD–QOLCT)

This section provides definitions of the left-, right-, and two-sided WVD–QOLCTs and shows its relationships.

The right-sided WVD–QOLCT is not equal to the left-sided one.

The left- and right-sided WVD–QOLCTs are not equal to the two-sided WVD–QOLCT. The proof of the lemma is obvious from the non-commutativity of quaternion algebra.

Despite non-equality, we can define the relationship between left- and right-sided WVD–QOLCTs, which is given in the next lemma.

Uncertainty Principles for the WVD–QOLCT

We prove the main results; first, we start with the logarithmic UP, it is a generalized variant of Heisenberg's UP, which is continued by Hardy's UP for the WVD–QOLCT.

Hardy's theorem is a qualitative UP, and it states that a function and its FT can't decrease rapidly simultaneously [80].

We establish Hardy's theorem related to the WVD–QOLCT.

Applications

In [92, 16, 93, 94–95, 96] different authors described an algorithm for the detection of LFM signals by hybrid transforms, that are more effective than the classical WVD.

We use the WVD–QOLCT as a tool for one-, and two-component LFM signal detection in the quaternion setting.

We state that the applications of the WVD–QOLCT to quaternion-valued LFM signal detection are effective and useful.

Conclusion
Some fundamental properties and several different forms of UPs associated with the 2D two-sided WVD–QOLCT are presented.

We apply the WVD–QOLCT to detect one-, and two-component quaternion-valued LFM signals.

The results of this paper analogously can be implemented for the 2D left-, and right-sided WVD–QOLCTs.

We will compare signal detection results of the WVD–QOLCT with existing transforms in our future work.

Acknowledgement
A machine generated summary based on the work of Urynbassarova, Didar; El Haoui, Youssef; Zhang, Feng

2022 in Circuits, Systems, and Signal Processing

2.4.4 Quaternionic Linear Canonical Wave Packet Transform

DOI: https://doi.org/10.1007/s00006-022-01224-0

Abstract-Summary
We introduce the notion of linear canonical wave packet transform in quaternionic settings and we name it the quaternionic linear canonical wave packet transform (QLCWPT).

Some key harmonic analysis results like energy conservation, inversion formula, characterization of range and some bounds of the QLCWPT are obtained.

Towards the culmination of this paper, we establish Heisenberg's uncertainty principle and the logarithmic uncertainty principle associated with the proposed transform (QLCWPT).

Introduction
The term wave packet transform (WPT) [97] is defined as the Fourier transform of a signal windowed with a given wavelet.

In 2015, he developed a theoretical linear algebra approach to the theory of wave packet transformations (WPT) over finite fields [98] that was the first of its kind.

The fractional wave packet transform and the conventional wave packet transform were later generalised by Prasad and Kundu [99] as linear canonical wave packet transform.

Since a few years, considerable attention has been paid to the representation of the linear canonical transform (LCT) for a quaternion valued signals, hence generating the quaternion linear canonical transform (QLCT).

The linear canonical wave packet transform (LCWPT) proposed by Prasad and Kundu [99] has not yet been applied to quaternions.

In this paper we will extend the concept of the LCWPT to quaternions, thus giving birth to the quaternion linear canonical wave packet transform (QLCWPT).

Quaternion Linear Canonical Wave Packet Transform
Due to the non commutative property of quaternion multiplication, the QLCWPT can be classified into three types as: two-sided QLCWPT, left-sided QLCWPT and right-sided QLCWPT.

We will use the two-sided QLCWPT to define the QLCWPT.

These properties play important roles in signal representation.

Uncertainty Principle for Quarternionic Linear Canonical Wave Packet Transform
One of the fundamentals of harmonic analysis and signal processing in general is the uncertainty principle [82].

Uncertainty principles have ramifications in two areas: quantum physics and signal processing.

The uncertainty principle is also known as the duration-bandwidth theorem in signal analysis because it asserts that a signal's width in the time domain (duration) and frequency domain (band-width) are limited in the sense that neither can be made arbitrarily thin.

In this uncertainty principle a lower bound was determined by the product of the effective widths of quaternion valued signals in the spatial and frequency domains.

Acknowledgement
A machine generated summary based on the work of Bhat, Younis Ahmad; Sheikh, N. A.

2022 in Advances in Applied Clifford Algebras

3 Other Quaternionic Integral Transforms (QIT)

Machine Generated Keywords Group, function, formula, Heisenberg, Gabor, theory, relation, finite, Clifford, give, QFT, theorem, pair, operator, vector field

3.1 Other QIT Theory

Machine Generated Keywords Group, function, Heisenberg, Gabor, theory, relation, finite, pair, operator, derivative, formula, transform quaternionic, Clifford, quaternionic, kernel

3.1.1 On the Quaternionic Short-Time Fourier and Segal–Bargmann Transforms

DOI: https://doi.org/10.1007/s00009-021-01745-1

Abstract-Summary
We study a special one-dimensional quaternion short-time Fourier transform (QSTFT).

Its construction is based on the slice hyperholomorphic Segal–Bargmann transform.

Extended:

It gives the possibility to write the 1D QSTFT using the reproducing kernel associated with Gabor space. Finally, we show that the 1D QSTFT follows a Lieb's uncertainty principle, some classical uncertainty principles for quaternionic linear operators in quaternionic Hilbert spaces were considered in [100].

Introduction
There has been an increased interest in the generalization of integral transforms to the quaternionic and Clifford settings.

Such kind of transforms are widely studied, since they help in the analysis of vector-valued signals and images.

A quaternionic short-time Fourier transform in dimension 2 is studied in [101].

In [102] the same transform is defined in a Clifford setting for even dimension of more than two.

We introduce an extension of the short-time Fourier transform in a quaternionic setting in dimension one.

This integral transform is used also in [103] to study some quaternionic Hilbert spaces of Cauchy–Fueter regular functions.

Preliminaries
A new approach to quaternionic regular functions was introduced and then extensively studied in several directions, and it is nowadays widely developed [104, 105, 106, 107].

This new theory contains polynomials and power series with quaternionic coefficients on the right, contrary to the Fueter theory of regular functions defined by means of the Cauchy-Riemann Fueter differential operator.

This connection holds in any odd dimension (and in the quaternionic case) and has been explained in [108] in the language of slice regular functions with values in the quaternions and slice monogenic functions with values in a Clifford algebra.

The theory of slice regular functions has several applications in operator theory and in Mathematical Physics.

Further Properties of the Quaternionic Segal–Bargmann Transform
We prove some new properties of the quaternionic Segal–Bargmann transform.

We start from the unitary property which is not found in the literature in explicit form.

1D Quaternion Fourier Transform
We study the one-dimensional quaternion Fourier transforms (QFT).

We are considering here the 1D left sided QFT studied in this chapter of the book [109].

Translation and Modulation: As in the classical case, we have a commutative relation between the two operators.

Quaternion Short-Time Fourier Transform with a Gaussian Window

The idea of the short-time Fourier transform is to obtain information about local properties of the signal f. In order to achieve this aim the signal f is restricted to an interval and then its Fourier transform is evaluated.

Fubini's theorem is combined with the Moyal formula for the QSTFT.

Through the 1D QSTFT we can prove in another way that the eigenfunctions of the 1D quaternion Fourier transform are given by the Hermite functions.

Acknowledgement

A machine generated summary based on the work of De Martino, Antonino; Diki, Kamal

2021 in Mediterranean Journal of Mathematics

3.1.2 The Quaternion Fourier Number Transform

DOI: https://doi.org/10.1007/s00034-018-0824-6

Abstract-Summary

We introduce the quaternion Fourier number transform (QFNT), which corresponds to a quaternionic version of the well-known number-theoretic transform.

We derive several theoretical results necessary to define the QFNT and investigate its main properties.

Differently from other quaternion transforms, which are defined over Hamilton's quaternions, the QFNT requires considering a quaternion algebra over a finite field.

Introduction

The definition of the QFNT is motivated by the increasing interest in mathematical tools and applications related to hypercomplex signal processing; in recent works, there have been proposed other new quaternion transforms [110, 111], algorithms concerning quaternion adaptive filters in the frequency domain [112], neural networks with quaternionic neurons [113, 114] and several quaternion-based techniques for color image processing [115, 116, 117, 118], just to mention a few.

The QFNT is a kind of number-theoretic counterpart of the (discrete) quaternion Fourier transform (QFT) [79, 109].

Depending on the field where the QFNT is established, it may still be possible to perform the multiplications necessary for its calculation by means of additions and bit-shift operations [119]; this makes the new transform attractive under the aspect of computational complexity, when compared to the QFT, and suggests that the algebraic structure where it is defined can be used as a surrogate field to perform error-free hypercomplex signal processing.

Generalized Quaternions

The general definition of these numbers is given as in [120, 121].

Operations and properties specifically related to quaternions over finite fields have not been clearly addressed in the literature.

At any case, some of these operations and properties are mere extensions of what one has for Hamilton's quaternions.

The ordinary Fourier number transform (FNT) is defined as in [119].

The Quaternion Fourier Number Transform

This plays an important role in the demonstration of the invertibility of the quaternion Fourier number transform.

We introduce a definition for the quaternion Fourier number transform and determine the respective inverse transform.

The transform corresponds to a finite field counterpart of the discrete quaternion Fourier transform [79, 109].

Properties of the QFNT

We develop some properties of the quaternion Fourier number transform.

To what happens with the ordinary Fourier number transform with respect to the discrete Fourier transform, such properties hold some analogy with those of the quaternion discrete Fourier transform.

We highlight the cyclic convolution property, which allows us to suppose that, also in the quaternionic context, it is possible to perform filtering in the transform domain employing modular arithmetic operations only, conveniently adapted to generalized quaternions over finite fields.

With the purpose of developing the QFNT cyclic convolution property, one considers the right definition of this transform.

Color Image Processing Using the QFNT

Of the possibility of computing the QFNT of color images, some more specific applications can be envisioned.

It may be possible to use the QFNT to create similar schemes directly applicable to color images like encryption of color images.

Even if there is no parametrization, it is possible that the application of the QFNT contributes to the robustness of image encryption schemes against statistical attacks.

In any case, considering the results presented the QFNT can be viewed as a potential candidate to be part of image encryption schemes, bringing the inherent possibility of processing in an aggregate way all layers of a color image.

If a non-quaternionic number transform is used (see, for example [122] and [123]), this is not possible, that is, the layers of a color image must be separately encrypted, as if they were independent monochromatic images.

Concluding Remarks

We conducted an investigation related to generalized quaternions over finite fields, identifying existing properties and deriving some new results in this context.

The QFNT seems to be suitable for applications in the scenario of image encryption; this is partially due to the fact that the QFNT is highly sensitive to changes in the data one desires to process, which is not true for the QFT defined over Hamilton quaternions.

In any case, our current investigations can be summarized as (i) introduction of new properties of generalized quaternions over finite fields, (ii) establishment of other properties of the QFNT, (iii) investigation of details related to the extension of the transform to two- and three-dimensional cases, (iv) definition of other types of quaternion number-theoretic transforms, such as cosine-, sine- and Hartley-type transforms and (v) investigation of specific QFNT applications.

Acknowledgement
A machine generated summary based on the work of da Silva, Luiz C.; Lima, Juliano B.

2018 in Circuits, Systems, and Signal Processing

3.1.3 The Quaternion Fourier Transform of Finite Measure and Its Properties

DOI: https://doi.org/10.1007/978-3-030-79606-8_6

Abstract-Summary
It is shown that according to the non-commutative nature of quaternion multiplications some properties of the Fourier transform of finite measure are not valid in the quaternion Fourier transform of finite measure.

Introduction
The quaternion Fourier [124, 79] is a non-trivial extension of the classical Fourier transform in the framework of quaternion algebra.

The quaternion Fourier transform of finite measure is intimately related to the quaternion Fourier transform and can be considered as a slight generalization of the quaternion Fourier transform.

The notation of finite measure in the framework of the quaternion Fourier transform and Clifford Fourier transform was introduced by the authors in [125, 126, 127], respectively.

Our first step is to recall the quaternion Fourier transform and its inversion formula, which is needed to define the quaternion Fourier transform of finite measures.

Several properties for the quaternion Fourier transform of finite measure are also discussed in this section.

Quaternion Fourier Transform and Bandlimited Functions
Let us start by recalling the definition of the quaternion Fourier transform and its inverse, which we will use later.

For a detailed exposition of its properties and applications, the reader may consult [128, 129, 130, 45, 61, 131, 132–133, 134].

Quaternion Fourier Transform of Finite Measure

Before we get into the details of the definition of the quaternion Fourier transform of finite measure and its essential properties, the following definition introduces weak convergence of measures.

The next result gives a sufficient condition of weak convergence in terms of the quaternion Fourier transform of finite measure.

Conclusion

The Fourier transform of finite measure has been extended to the quaternion plane called the quaternion Fourier transform of finite measure.

Its properties are generalizations of corresponding properties of the Fourier transform of finite measure in real and complex regions.

We have shown that some properties of the Fourier transform of finite measure do not hold in the setting of the quaternion Fourier transform of finite measure.

Acknowledgement

A machine generated summary based on the work of Bahri, Mawardi; Rahim, Amran; Nur, Muh.; Amir, Amir Kamal

2021 in Studies in Systems, Decision and Control

3.1.4 The 2-D Hyper-Complex Gabor Quadratic-Phase Fourier Transform and Uncertainty Principles

DOI: https://doi.org/10.1007/s41478-022-00445-7

Abstract-Summary

We present a novel integral transform known as the 2-D Hyper-complex (quaternion) Gabor quadratic-phase Fourier transform (Q-GQPFT), which is an embodiment of several well known signal processing tools.

Introduction

To overcome a drawback we used the Q-QPFT to generate a new transform called the 2-D Hyper-complex (quaternion) Gabor quadratic-phase Fourier transform (Q-GQPFT).

Keeping in view the contemporary trends in time-frequency analysis, it is both theoretically interesting and practically useful to propose a generalized quaternion Gabor quadratic phase Fourier transform that can efficiently localize the quadratic-phase spectrum of a non-transient quaternion signal in the time-frequency plane.

The main purpose of this paper is to rigorously study the 2-D Hyper-complex (quaternion) Gabor quadratic phase Fourier transform.

The highlights of this study are. To propose the definition of a novel 2-D Hyper-complex (Quaternion) Qudratic-phase Fourier Transform (Q-QPFT). To propose the definition of novel integral transform named the 2-D Hyper-complex (quaternion) quadratic-phase Fourier transform (Q-QPFT). To establish the Heisenberg and logarithmic uncertainty inequalities associated with the 2-D Hyper-complex (quaternion) quadratic-phase Fourier transform (Q-QPFT).

2-D Hyper-Complex Fourier Transform and 2-D Hyper-Complex Quadratic-Phase Fourier Transform

We will introduce the definition of the novel 2-D Hyper-complex (Quaternion) Qudratic-phase Fourier Transform (Q-QPFT) which is a generalization of the classical Qudratic-phase Fourier transform [135].

Because of the non-commutative property of quaternion multiplication, there are three different types of the Q-QPFT: the left-sided Q-QPFT, the right-sided Q-QPFT, and the two-sided Q-QPFT.

We mainly focus on the two-sided Q-QPFT. (Q-QPFT).

2-D Hyper-Complex Gabor Qudratic-Phase Fourier Transform

We formally introduce the notion of novel 2-D Hyper-complex (Quaternion) Gabor Quadratic-Phase Fourier transforms (Q-GQPFTs).

We shall establish the relation between Quaternion Gabor Quadratic-Phase Fourier Transforms and Quaternion Quadratic-Phase Fourier Transforms and then investigate several basic properties of (two-sided) Q-GQPFT which are important for signal representation in signal processing.

These properties are vital in signal processing.

Uncertainty Principles for Quaternion Q-GQPFT

Heisenberg's uncertainty principle lies at the heart of any time-frequency transform, as it enables us to detect the optimal simultaneous resolution in time and frequency domains.

The uncertainty principles for the linear canonical transform, the windowed linear canonical transform and their counter parts in the quaternion domain have been discussed in Refs. [136, 137, 31, 138, 74, 139–140].

The authors have established the uncertainty principles associated with the quadratic-phase Fourier transform and short time quadratic-phase Fourier transform in Refs. [141, 142].

Conclusion

We defined the 2-D Hyper-complex (quaternion) quadratic-phase Fourier transform (Q-QPFT) and then proposed the definition of novel Q-GQPFT, which is a modified version of the classical windowed quadratic-phase Fourier transform to quaternion-valued signals.

We studied various properties of the proposed Q-GQPFT, including Moyal's formula, a reconstruction formula, isometry and a reproducing kernel formula.

Acknowledgement

A machine generated summary based on the work of Bhat, M. Younus; Dar, Aamir H. 2022 in The Journal of Analysis

3.1.5 Monogenic Cauchy Implies Holomorphic Bochner–Martinelli

DOI: https://doi.org/10.1007/s00006-022-01213-3

Introduction

After the quaternions had been invented by Hamilton, there arose a variety of function theories of quaternions which are similar to the theory of holomorphic functions of a complex variable.

This theory has been further developed into the monogenic theory of Clifford algebra valued functions or spinor valued functions [143, 144, 145].

In Fueter's theory, a very important strategy was devised to construct monogenic functions in terms of holomorphic functions (see [146] for example).

This builds a strong connection between holomorphic and monogenic functions.

We investigate a system of two equations. Such a solution is called the quaternion-valued holomorphic function.

This means that the limit exists so that f is a linear function for some constant quaternions a, b. Together with the observations above, we conclude that the relations among the linear polynomial, the holomorphic function, and the monogenic function can be described in terms of a set of subsystems of equations, given by Cauchy–Fueter type operators.

Inner Product and Chain Rules in Quaternions

We require a quaternionic version of the chain rules, which is of interest by its own.

We remark on the precise version of the chain rules. Notice that the order matters for functions in the formula.

It is sufficient to verify the chain rule at any given point.

Quaternion-Valued Holomorphic Functions

A quaternion-valued holomorphic function can be regarded as a holomorphic map and vice versa.

We point out that these two sets have no inclusive relations.

To exhibit the relation between holomorphic and monogenic, we can interpret the holomorphic version of Cartan's uniqueness theorem (see [147]) in terms of monogenic functions.

Acknowledgement

A machine generated summary based on the work of Li, Yong; Ren, Guangbin 2022 in Advances in Applied Clifford Algebras

3.1.6 Boundedness and Uniqueness of Quaternion Weyl Transform

DOI: https://doi.org/10.1007/s11868-022-00454-x

Introduction
The boundedness property of the Weyl transform has been considered in different setups including the Heisenberg group, the quaternion Heisenberg the upper, upper half-plane, the Euclidean and Heisenberg motion groups [148, 149, 150, 151].

The uncertainty principle for the quaternion Fourier transform (QFT) has received significant attention [78, 45, 49, 133].

The non-commutativity of the quaternion multiplication and the Fourier kernel make the QFT different from the classical Fourier transform.

The QFT has an important application in data analysis, particularly in color image processing, etc. Since the quaternion algebra decomposes into two complex planes, the QFT can be split into two Euclidean Fourier transforms, which makes the QFT accessible.

Preliminaries and Auxiliary Results
The quaternion algebra was discovered by W. R. Hamilton in 1843.

We have a version of Hardy's theorem associated with the QFT.

The Fourier–Wigner Transform and Weyl Transform
Throughout the section, we use the terminology Fourier–Wigner transform, Wigner transform and Weyl transform instead of quaternion Fourier–Wigner transform, quaternion Wigner transform and quaternion Weyl transform, respectively, since it will be clear from the context.

The zero set of the Wigner transform is useful in studying the injectivity of a general Berezin transform and the generalized Berezin quantization problem.

In [152], the authors studied under which conditions the Euclidean Wigner transform never vanishes.

It is shown that when f and g are generalized Gaussian, then the Euclidean Wigner transform is never vanishing.

Uniqueness Results
The following is the main result of the section.

The version of Beurling's theorem for the Fourier-Weyl transform on step two nilpotent Lie groups proved in [153] can be generalized for the quaternion Fourier-Weyl transform.

In a remarkable result, Helgason proved a support theorem (see [154]).

Acknowledgement
A machine generated summary based on the work of Dalai, Rupak K.; Ghosh, Somnath; Srivastava, R. K.

2022 in Journal of Pseudo-Differential Operators and Applications

3.1.7 Slice Regular Functions in Several Variables

DOI: https://doi.org/10.1007/s00209-022-03066-9

Abstract-Summary
One of the relevant aspects of the theory is the validity of a Cauchy-type integral formula and the existence of ordered power series expansions.

The theory includes all polynomials and power series with ordered variables and right coefficients in the algebra.

We study the real dimension of the zero set of polynomials in the quaternionic and octonionic cases and give some results about the zero set of polynomials with Clifford coefficients.

Introduction
We then prove the smoothness properties of slice functions.

All polynomials (with ordered monomials) turn out to be slice regular functions.

We show that slice regularity in several variables has an interpretation, by means of the spherical value and spherical derivatives, in terms of slice regularity in one variable.

We investigate Leibniz's rule and we prove the stability of slice regularity under the so-called slice tensor product of slice functions.

We define a slice Cauchy kernel associated to any given slice regular function, and obtain a Cauchy integral formula.

As we will see later, the concepts of spherical value and spherical derivative in one variable will have a central role to get a characterization of slice regularity in several variables in terms of separate one variable regularity.

We refer the reader to [155, §§3,4] for more properties of slice functions and slice regularity in one variable.

Slice Functions
We are in a position to introduce the notion of slice function in several variables.

A by-product of the proof of the mentioned proposition reads as follows.

The pointwise product of two stem functions is still a stem function.

The next result concerns the relation between slice tensor and pointwise products, with F and G being the stem functions inducing f and g, respectively.

Slice Regular Functions
The next result contains some characterizations of slice regularity.

We specialize this definition to the slice regular case.

We say that g is a slice regular function.

F is slice regular if and only if f is a slice regular function.

We assume that f is slice regular.

We say that a function s is a sum of the convergent power series s(X).

Acknowledgement
A machine generated summary based on the work of Ghiloni, Riccardo; Perotti, Alessandro

2022 in Mathematische Zeitschrift

3.1.8 Bounded Spherical Functions and Heisenberg Inequality on Quaternionic Heisenberg Group

DOI: https://doi.org/10.1007/s43036-022-00188-z

Introduction

The uncertainty principles play an important role in harmonic analysis, and they state that it is impossible for a non-zero function and its Fourier transform to be simultaneously small.

The most common quantitative formulation of the uncertainty principle is the Heisenberg–Pauli–Weyl inequality [82, 156].

This group plays an important role in several branches of mathematics such as harmonic analysis, representation theory, partial differential equations, and quantum mechanics.

In their article [157], the authors have determined the bounded spherical functions associated with the Gelfand pair formed by the quaternionic Heisenberg group and the subgroup K of the quaternions of module 1, and then, in this paper, we will construct the bounded spherical functions by a different method to the one given in [157] and we find an appropriate definition of the spherical Gabor transform on the Quaternionic Heisenberg group.

Acknowledgement

A machine generated summary based on the work of Amghar, Walid; Fahlaoui, Said 2022 in Advances in Operator Theory

3.2 *Other QIT Applications*

Machine Generated Keywords Vector field, Hitzer, sample, sample formula, vector, scalar, convolution, prolate, prolate spheroidal, spheroidal, spheroidal wave, wave, property two-side, particular, datum

3.2.1 Two-Sided Fractional Quaternion Fourier Transform and Its Application

DOI: https://doi.org/10.1186/s13660-021-02654-3

Abstract-Summary

We introduce the two-sided fractional quaternion Fourier transform (FrQFT) and give some properties of it.

We give an example to illustrate the application of the two-sided FrQFT and its inverse transform in solving partial differential equations.

Introduction

The quaternion Fourier transform (QFT) is one of the generalized forms of the classical Fourier transform in high dimensional space and has been proved to be very useful in signal processing, non-marginal color image processing, electromagnetism, multi-channel processing, quantum mechanics, and partial differential systems.

Some properties of the QFT and the two-sided QFT have been studied [158–159].

Hitzer [131] researched the QFT properties useful for applications to differential equations, image processing and optimized numerical implementations and studied the general linear transformation behavior of the QFT with matrices.

In [160], Bahri proposed the uncertainty principle for the two-sided QFT.

That uncertainty principle described that the spread of a quaternion-valued function and its two-sided QFT are inversely proportional.

Based on the nature of the two-sided QFT, we study the relationship between the two-sided QFT and the two-sided FrQFT.

Some Properties of the Two-Sided FrQFT

We state some properties of the two-sided FrQFT.

We first give a definition of the two-sided FrQFT and its inverse transformation.

Next, we give some important properties of the two-sided FrQFT; we begin with the shift property.

We can see that the two-sided FrQFT has norm-preserving properties.

The Application of the Two-Sided Fractional QFT

We give an application of differential properties of the two-sided FrQFT in solving partial differential equations.

We find solutions to partial differential equations.

Conclusion

Using the basic concepts of quaternion algebra we introduced a two-sided FrQFT.

Important properties of the two-sided FrQFT such as shift, differential properties, and Parseval identities were demonstrated.

We mention that applications of the two-sided FrQFT in signal processing, non-marginal color image processing and electromagnetism, etc., are not given.

Acknowledgement

A machine generated summary based on the work of Li, Zunfeng; Shi, Haipan; Qiao, Yuying

2021 in Journal of Inequalities and Applications

3.2.2 Analyzing Real Vector Fields with Clifford Convolution and Clifford–Fourier Transform

DOI: https://doi.org/10.1007/978-1-84996-108-0_7

Abstract-Summary
While image processing of scalar data is a well-established discipline, there is a lack of similar methods for vector data.

This paper surveys a particular approach defining convolution operators on vector fields using geometric algebra.

A comparison is tried with related approaches for a Fourier transform of spatial vector or multivector data.

Acknowledgement
A machine generated summary based on the work of Reich, Wieland; Scheuermann, Gerik

2010 Geometric algebra computing in engineering and computer science.

3.2.3 Novel Sampling Formulas Associated with Quaternionic Prolate Spheroidal Wave Functions

DOI: https://doi.org/10.1007/s00006-017-0815-x

Abstract-Summary
The Whittaker–Shannon–Kotel'nikov (WSK) sampling theorem provides a reconstruction formula for bandlimited signals.

A generalization is employed to obtain novel sampling formulas for bandlimited quaternion-valued signals.

A special case of our result shows that the 2D generalized prolate spheroidal wave signals obtained by Slepian can be used to achieve a sampling series of cube-bandlimited signals.

Acknowledgement
A machine generated summary based on the work of Cheng, Dong; Kou, Kit Ian

2017 in Advances in Applied Clifford Algebras

References

1. Bahri, M., Ashino, R., Vaillancourt, R.: Convolution Theorems for Quaternion Fourier Transform: Properties and Applications. Abstract and Applied Analysis, vol. 2013. Hindawi Publishing Corporation, London (2013)
2. Laugesen, R.S.S., Weaver, N., Weiss, G.L., Wilson, E.N.: A characterization of the higher dimensional groups associated with continuous wavelets. J. Geom. Anal. 12(1), 89–102 (2002)
3. Dahlke, S., Kutyniok, G., Maass, P., Sagiv, C., Stark, H.-G., Teschke, G.: The uncertainty principle associated with the continuous Shearlet transform. Int. J. Wavelets Multiresolut. Inf. Process. 6, 157–181 (2008)
4. Dahlke, S., Steidl, G., Teschke, G.: The continuous shearlet transform in arbitrary space dimensions. J. Fourier Anal. Appl. 16, 340–364 (2010)
5. Guo, K., Labate, D.: Characterization and analysis of edges using the continuous shearlet transform. SIAM J. Imaging Sci. 2, 959–986 (2009)

6. Guo, K., Labate, D.: Characterization of piecewise-smooth surfaces using the 3D continuous shearlet transform. J. Fourier Anal. Appl. 18, 488–516 (2012)
7. Guo, K., Labate, D.: Analysis and identification of multidimensional singularities using the continuous shearlet transform. In: Shearlet. Birkhäuser, Boston, pp. 69–103 (2012)
8. Kutyniok, G., Labate, D.: Resolution of the Wavefront Set using continuous Shearlets. Trans. Amer. Math. Soc. 361, 2719–2754 (2009)
9. Kutyniok, G., Labate, D.: Introduction to shearlets. In: Shearlet. Birkhäuser, Boston, pp. 1–38 (2012)
10. Liu, S., Hu, S., Xiao, Y., An, L.: A Bayesian shearlet shrinkage for SAR image denoising via sparse representation. Multidim. Syst. Sign Process. 25, 683–701 (2014)
11. Guo, K., Kutyniok, G., Labate, D.: Sparse multidimensional representations using anisotropic dilation and shear operators. In: Chen, G., Lai, M.J. (eds.) Wavelets and Splines, pp. 189–201. Athens (2005)
12. Heisenberg, W.: Uber den anschaulichen inhalt der quanten theoretischen kinematik und mechanik. Zeitschrift für Physik 43, 172–198 (1927)
13. Kennard, E.H.: Zur Quantenmechanik einfacher Bewegungstypen. Z. Phys. 44, 326–352 (1927)
14. Weyl, H.: Gruppentheorie und Quantenmechanik, S. Hirzel, Leipzig. Revised English edition: Groups and Quantum Mechanics, Dover (1950)
15. Guo, C., Ma, Q., Zhang, L.: Spatio-temporal saliency detection using phase spectrum of quaternion Fourier transform. In: 2008 IEEE Conference on Computer Vision Pattern Recognition, pp. 1–8 (2008)
16. Fan, X.L., Kou, K.I., Liu, M.S.: Quaternion Wigner–Ville distribution associated with the linear canonical transforms. Signal Process. 130, 129–141 (2017)
17. Fu, Y.X., Li, L.Q.: Generalized analytic signal associated with linear canonical transform. Opt. Commun. 281, 1468–1472 (2008)
18. Xu, G., Wang, X., Xu, X.: Generalized Hilbert transform and its properties in 2D LCT domain. Signal Processing 89, 1395–1402 (2009)
19. Kou, K.I., Liu, M.S., Morais, J.P., et al.: Envelope detection using generalized analytic signal in 2D QLCT domains. Multidimens. Syst. Signal Process. 28(4), 1343–1366 (2017)
20. Kou, K.I., Ou, J.Y., Morais, J.: On uncertainty principle for quaternionic linear canonical transform, Abstract and Applied Analysis Vol. 2013, Article ID 725952, 14 pages
21. Pei, S.C., Ding, J.J., Huang, J.D., Guo, G.C.: Short response Hilbert transform for edge detection, IEEE (2008). IEEE Xplore Digital Library (2008). https://ieeexplore.ieee.org/document/4746029
22. Mustard D.: Uncertainty principle invariant under fractional Fourier transform. J. Aust. Math. Soc. Ser. B 33, 180–191 (1991)
23. Ozaktas, H.M., Kutay, M.A., Zalevsky, Z.: The Fractional Fourier Transform with Applications in Optics and Signal Processing. Wiley, New York (2000)
24. Collins, S.A.: Lens-system diffraction integral written in terms of matrix optics. J. Opt. Soc. Am. 60, 1168–1177 (1970)
25. Moshinsky, M., Quesne, C.: Linear canonical transformations and their unitary representations. J. Math. Phys. 12(8), 1772–1780 (1971)
26. Hitzer E, Sangwine S J. The Orthogonal 2D Planes Split of Quaternions and Steerable Quaternion Fourier Transformations//Hitzer E, Sangwine S. Quaternion and Clifford Fourier Transforms and Wavelets. Trends in Mathematics. Basel: Birkhüauser, 2013
27. Bahri, M., Ashino, R., Vaillancourt, R.: Continuous quaternion Fourier and wavelet transforms. Int. J. Wavelets Multiresolut. Inf. Process. 12, 1460003 (2014)
28. Akila, L., Roopkumar, R.: Ridgelet transform on quaternion valued functions. Int. J. Wavelets Multiresolut. Inf. Process. 14, 1650006 (2016)
29. Akila, L., Roopkumar, R.: Quaternionic Stockwell transform. Integ. Transf. Spec. Funct. 27(6), 484–504 (2016)
30. Akila, L., Roopkumar, R.: Quaternionic curvelet transform. Optik. 131, 255–266 (2017)

31. Gao, W.B., Li, B.Z.: Quaternion windowed linear canonical transform of two-dimensional signals. Adv. Appl. Clifford Algebras. 30, (2020). https://doi.org/10.1007/s00006-020-1042-4
32. Achak, A., Bouhlal, A., Daher, R., et al.: Titchmarsh's theorem and some remarks concerning the right-sided quaternion Fourier transform. Bol. Soc. Mat. Mex. 26, 599–616 (2020)
33. Achak, A., Abouelaz, A., Daher, R., Safouane, N.: Uncertainty principles for the quaternion linear canonical transform. Adv. Appl. Clifford Algebras 29(5), 1–19 (2019)
34. Ahmad, O., Sheikh, N.A.: Novel special affine wavelet transform and associated uncertainty inequalities. Int. J. Geom. Methods Mod. Phys. 18(4), 2150055 (16 pages) (2021)
35. Bahri, M., Ashino, R.: Two-dimensional quaternion linear canonical transform: properties, convolution, correlation, and uncertainty principle. Hindawi J. Math. 13, 1062979 (2019)
36. Brahim, K., Tefjeni, E.: Uncertainty principle for the two-sided quaternion windowed Fourier transform. Integral Transf. Spec. Funct. 30(5), 362–382 (2019)
37. Brahim, K., Tefjeni, E.: Uncertainty principle for the two-sided quaternion windowed Fourier transform. J. Pseudo Differ. Oper. Appl. (2019). https://doi.org/10.1007/s11868-019-00283-5
38. Iwo B.B.: Formulation of the uncertainty relations in terms of the Rényi entropies. Phys. Rev. A 74, 052101 (2006)
39. Maassen, H.: A Discrete Entropic Uncertainty Relation, Quantum Probability and Applications, V, pp. 263–266. Springer, New York (1988)
40. Donoho, D.L., Stark, P.B.: Uncertainty principles and signal recovery. J. Appl. Math. 49(3), 906–931 (1989)
41. Shinde S., Gadre V.M.: An uncertainty principle for real signals in the fractional Fourier transform domain. IEEE Trans. Signal Process. 49(11), 2545–2548 (2001)
42. Xu, G.L., Wang, X.T., Xu, X.G.: The logarithmic, Heisenberg's and short-time uncertainty principles associated with fractional Fourier transform. Signal Process. 89(3), 339–343 (2009)
43. Zhang, Y.N., Li, B.Z.: Generalized uncertainty principles for the two-sided quaternion linear canonical transform. In: Proceedings of the IEEE international conference on acoustics speech and signal processing, ICASSP, pp. 4594–4598 (2018)
44. S.C. Pei, J.J. Ding, J.H. Chang, Efficient implementation of quaternion Fourier transform, convolution, and correlation by 2-D complex FFT. IEEE Trans. Signal Process. 49 (11), 2783–2797 (2001)
45. Chen L-P, Kou K.-I., Liu M.-S.: Pitt's inequality and the uncertainty principle associated with the quaternion Fourier transform. J. Math. Anal. Appl. 423, 681–700 (2015)
46. Lian, P.: Uncertainty principle for the quaternion Fourier transform. J. Math. Anal. Appl. (2018). https://doi.org/10.1016/j.jmaa.2018.08.002
47. Y.N. Zhang, B.Z. Li, Generalized uncertainty principles for the two-sided quaternion linear canonical transform, in Proceedings of the IEEE International Conference on Acoustics Speech and Signal Processing, ICASSP, pp. 4594–4598 (2018). https://doi.org/10.1109/ICASSP.2018.8461536
48. El Haoui, Y., Fahlaoui, S.: The uncertainty principle for the two-sided quaternion Fourier transform. Mediterr. J. Math. (2017). https://doi.org/10.1007/s00009-017-1024-5
49. Y. El Haoui, S. Fahlaoui, Beurling's theorem for the quaternion Fourier transform. J. Pseudo-Differ. Oper. Appl. (2019). https://doi.org/10.1007/s11868-019-00281-7
50. M. Bahri, R. Ashino, Simplified proof of uncertainty principle for quaternion linear canonical transform, in Abstract and Applied Analysis. (Hindawi, London, 2016), pp. 1–11
51. B. Hu, Y. Zhou, L.D. Lie, J.Y. Zhang, Polar linear canonical transformin quaternion domain. J. Inf. Hiding Multimed. Signal Process. 6(6), 1185–1193 (2015)
52. K.I. Kou, J. Morais, Asymptotic behaviour of the quaternion linear canonical transform and the Bochner-Minlos theorem. Appl. Math. Comput. 247(15), 675–688 (2014)
53. K.I. Kou, J. Morais, Y. Zhang, Generalized prolate spheroidal wave functions for offset linear canonical transform in Clifford analysis. Math. Methods Appl. Sci. 36(9), 1028–1041 (2013)
54. K.I. Kou, M. Liu, S. Tao, Herglotz's theorem and quaternion series of positive term. Math. Methods Appl. Sci. 39(18), 5607–5618 (2016)

55. K.I. Kou, J.Y. Ou, J. Morais, Uncertainty principles associated with quaternionic linear canonical transforms. Math. Meth. Appl. Sci. 39(10), 2722–2736 (2016)
56. Y.N. Zhang, B.Z. Li, Novel uncertainty principles for two-sided quaternion linear canonical transform. Adv. Appl. Clifford Algebr. 28(1), 1–15 (2018)
57. K.I. Kou, R.H. Xu, Y.H. Zhang, Paley-Wiener theorems and uncertainty principles for the windowed linear canonical transform. Math. Methods Appl. Sci. 35(17), 2122–2132 (2012)
58. L. Huang, K. Zhang, Y. Chai, S.Q. Xu, Uncertainty principle and orthogonal condition for the short-time linear canonical transform. Signal Image Video Process. 10, 1177–1181 (2016)
59. Z. Xiong, Y. X. Fu, A kind of windowed quaternionic linear canonical transform. Master Thesis, Hubei University, Wuhan, China, (2015)
60. E. Hitzet, S.J. Sungwine, Quaternion and Clifford Fourier Transforms and Wavelets (Birkhäuser, Basel, 2013)
61. E.M.S. Hitzer, Quaternion Fourier transform on quaternion fields and generalizations. Adv. Appl. Clifford Algebr. 17(3), 497–517 (2007)
62. L.P. Chen, K.I. Kou, M.S. Liu, Pitt's inequality and the uncertainty principle associated with the quaternion Fourier transform. J. Math. Anal. Appl. 423(1), 681–700 (2015)
63. A. Achak, A. Abouelaz, R. Daher, N. Safouane, Uncertainty principles for the quaternion linear canonical transform. Adv. Appl. Clifford Algebr. 29(5), 1–19 (2019)
64. Chen, Q., Qian, T.: Sampling theorem and multi-scale spectrum based on non-linear Fourier atoms. Appl. Anal. 88(6), 903–919 (2009)
65. Chen, Q., Wang, Y., Wang, Y.: A sampling theorem for non-bandlimited signals using generalized sinc functions. Comput. Math. Appl. 56(6), 1650–1661 (2008)
66. Liu, Y.L., Kou, K.I., Ho, I.T.: New sampling formulae for non-bandlimited signals associated with linear canonical transform and nonlinear Fourier atoms. Signal Process. 90(3), 933–945 (2010)
67. Cheng, D., Kou, K.I.: Novel sampling formulas associated with quaternionic prolate spheroidal wave functions. Adv. Appl. Clifford Algebras 27(4), 2961–2983 (2017)
68. Cheng, D., Kou, K.I.: Generalized sampling expansions associated with quaternion Fourier transform. Math. Methods Appl. Sci. 41(11), 4021–4032 (2018)
69. Hu, X., Cheng, D., Kou, K.: Sampling formulas for 2D quaternionic signals associated with various quaternion Fourier and linear canonical transforms. Front. Inf. Technol. Electr. Eng. (2021)
70. Xiao-xiao, H., Kou, K.I.: Inversion theorems of quaternion Fourier and linear canonical transforms. Math. Methods Appl. Sci. 40(7), 2421–2440 (2017)
71. Xiang, M., Dees, B.S., Mandic, D.P.: Multiple-model adaptive estimation for 3-D and 4-D signals: a widely linear quaternion approach. IEEE Trans. Neural Netw. Learn. Syst. 30(1), 72–84 (2019)
72. Chen, Y., Xiao, X., Zhou, Y.: Low-rank quaternion approximation for color image processing. IEEE Trans. Image Process. 29, 1426–1439 (2020)
73. Kou, K., Morais, J., & Zhang, Y. (2013). Generalized prolate spheroidal wave functions for offset linear canonical transform in Clifford analysis. Mathematical Methods in the Applied Sciences, 36, 1028–1041.
74. Yang, Y., Kou, K.I.: Uncertainty principles for hypercomplex signals in the linear canonical transform domains. Signal Process. 95, 67–75 (2014)
75. B. Mawardi, E. Hitzer, A. Hayashi, and R. Ashino. An uncertainty principle for quaternion Fourier transform. Computers and Mathematics with Applications, 56(9):2411– 2417, 2008.
76. Mawardi, B., Hitzer, E., Ashino, R., & Vaillancourt, R. (2010). Windowed Fourier transform for two-dimensional quaternionic signals. Applied Mathematics and Computation, 216, 2366–2379.
77. Yin, M., Liu, W., Shui, J., & Wu, J. (2012). Quaternion wavelet analysis and application in image denoising. Mathematical Problems in Engineering, 2012, 493976. https://doi.org/10.1155/2012/493976.

78. Bulow T, Sommer G, 2001. Hypercomplex signals—a novel extension of the analytic signal to the multidimensional case. IEEE Trans Signal Process, 49(11):2844–2852. https://doi.org/10.1109/78.960432
79. Ell T.A., Sangwine S.J.: Hypercomplex Fourier transforms of color images. IEEE Trans. Image Process. 16(1), 22–35 (2007)
80. El Haoui, Y., Hitzer, E.: Generalized uncertainty principles associated with the quaternionic offset linear canonical transform. Complex Var. Elliptic Equ. (2021). https://doi.org/10.1080/17476933.2021.1916919
81. Zhu, X., Zheng, S.: Uncertainty principles for the two-sided offset quaternion linear canonical transform. Math. Method Appl. Sci. 44, 14236–14255 (2021)
82. Folland, G.B., Sitaram, A.: The uncertainty principle: a mathematical survey. J. Fourier Anal. Appl. 3(3), 207–238 (1997)
83. Beckner, W.: Pitt's inequality and the uncertainty principle. Proc. Am. Math. Soc. 123, 1897–1905 (1995)
84. Shah, F.A., Teali, A.A., Tantary, A.Y.: Windowed special affine Fourier transform. J. Pseudo-Differ. Oper. Appl. 11, 13891420 (2020)
85. Wilczok, E.: New uncertainty principles for the continuous Gabor transform and the continuous wavelet transform. Doc. Math. 5, 201–226 (2000)
86. Srivastava, H.M., Kumar, M., Pradhan, T.: A framework of linear canonical Hankel transform pairs in distribution spaces and their applications. Rev. Real Acad. Cienc. Exactas Fís. Natur. Ser. A Mat. 115, 171 (2021)
87. Srivastava, H.M., Mohammed, P.O., Guirao, J.L.G., Hamed, Y.S.: Link theorem and distributions of solutions to uncertain Liouville-Caputo difference equations. Discrete Continuous Dyn. Syst. S. 15, 427–440 (2022)
88. Srivastava, H.M., Shah, F.A., Garg, T.K., Lone, W.Z., Qadri, H.L.: Non-separable linear canonical wavelet transform. Symmetry 13, 2182 (2021)
89. Y. El Haoui, Erratum to: The Wigner–Ville distribution associated with the quaternion offset linear canonical transform. Anal. Math. 48, 279–282 (2022). https://doi.org/10.1007/s10476-021-0107-5
90. M. El Kassimi, Y. El Haoui, S. Fahlaoui, The Wigner–Ville distribution associated with the quaternion offset linear canonical transform. Anal. Math. 45, 787–802 (2019). https://doi.org/10.1007/s10476-019-0007-0
91. Hahn, S.L., Snopek, K.M.: Wigner distributions and ambiguity function of 2-D quaternionic and monogenic signals. IEEE Trans. Sigal Process. 53(8), 3111–3128 (2005)
92. R.F. Bai, B.Z. Li, Q.Y. Cheng, Wigner–Ville distribution associated with the linear canonical transform. J. Appl. Math. 2012, 1–14 (2012). https://doi.org/10.1155/2012/740161
93. Y.E. Song, X.Y. Zhang, C.H. Shang, H.X. Bu, X.Y. Wang, Wigner–Ville distribution based on the linear canonical transform and its applications for QFM signal parameters estimation. J. Appl. Math. 2014, 1–8 (2014). https://doi.org/10.1155/2014/516457
94. Urynbassarova, D., Li, B.-Z., Tao, R.: Convolution and correlation theorems for Wigner-Ville distribution associated with the offset linear canonical transform. Optik 157, 455–466 (2018)
95. D. Urynbassarova, B.Z. Li, R. Tao, The Wigner–Ville distribution in the linear canonical transform domain. IAENG Int. J. Appl. Math. 46, 559–563 (2016)
96. Z.C. Zhang, Unified Wigner–Ville distribution and ambiguity function in the linear canonical transform domain. Signal Process. 114, 45–60 (2015). https://doi.org/10.1016/j.sigpro.2015.02.016
97. Posch, T.E.: The wave packet transform (WPT) as applied to signal processing. In: Proceedings of the IEEE-SP International Symposium on Time-Frequency and Time-Scale Analysis, pp. 143–146 (1992). https://doi.org/10.1109/TFTSA.1992.274216
98. Ghaani Farashahi, A.: Wave packet transform over finite fields. Electron. J. Linear Algebra 30, 507–529 (2015). https://doi.org/10.13001/1081-3810.2903
99. Prasad, A., Kundu, M.: Linear canonical wave packet transform. In: Integral Transforms and Special Functions, pp. 1–19 (2021). https://doi.org/10.1080/10652469.2020.1867128

100. Xu, Z., Ren, G.: Sharper uncertainty principles in quaternionic Hilbert spaces. Math. Methods Appl. Sci. 43, 1608–1630 (2020)
101. Bahri, M., Ashino, R.: Two-Dimensional Quaternionic Window Fourier Transform, in Fourier Transform—Approach to Scientific Principles. InTechOpen, London (2011)
102. De Martino, A.: On the Clifford short-time Fourier transform and its properties (in preparation)
103. Diki, K., Krausshar, R.S., Sabadini, I.: On the Bargmann–Fock–Fueter and Bergman–Fueter integral transform. J. Math. Phys. 60, 1–26 (2019)
104. Alpay, D., Colombo, F., Sabadini, I.: Slice Hyperholomorphic Schur Analysis Operator. Theory: Advances and Applications, vol. 256. Birkhäuser, Basel (2017)
105. Colombo, F., Sabadini, I., Struppa, D.C.: Noncommutative Functional Calculus, Progress in Mathematics, vol. 289. Birkhäuser/Springer Basel AG, Basel (2011)
106. Colombo, F., Sabadini, I., Struppa, D.C.: Entire Slice Regular Functions. SpringerBriefs in Mathematics. Springer, Cham (2016)
107. Gentili G, Stoppato C, Struppa D C. Regular functions of a quaternionic variable. Heidelberg: Springer, 2013
108. Colombo, F., Sabadini, I., Sommen, F.: The Fueter mapping theorem in integral form and the F-functional calculus. Math. Methods Appl. Sci. 33, 2050–2066 (2010)
109. Ell, T.A., Bihan, N.L., Sangwine, S.J.: Quaternion Fourier Transforms for Signal and Image Processing. Wiley, New York (2014)
110. Fletcher, P., Sangwine, S.J.: The development of the quaternion wavelet transform. Sig. Process. 136, 2–15 (2017). https://doi.org/10.1016/j.sigpro.2016.12.025
111. Roopkumar, R. Quaternionic one-dimensional fractional Fourier transform. Optik 127(24):11657–11661 (2016)
112. F. Ortolani, D. Comminiello, M. Scarpiniti, A. Uncini, Frequency domain quaternion adaptive filters: algorithms and convergence performance. Signal Process. 136(7), 69–80 (2017)
113. M. Kobayashi, Fixed points of split quaternionic hopfield neural networks. Signal Process. 136(7), 38–42 (2017)
114. T. Minemoto, T. Isokawa, H. Nishimura, N. Matsui, Feed forward neural network with random quaternionic neurons. Signal Process. 136(7), 59–68 (2017)
115. B. Augereau, P. Carré, Hypercomplex polynomial wavelet-filter bank transform for color image. Signal Process. 136(7), 16–28 (2017)
116. R. Lan, Y. Zhou, Quaternion-Michelson descriptor for color image classification. IEEE Trans. Image Process. 25(11), 5281–5292 (2016)
117. Liu, C., Li, J., Fu, B.: Magnitude-phase of quaternion wavelet transform for texture representation using multilevel copula. IEEE Sig. Process. Lett. 20(8), 799–802 (2013)
118. Y. Xu, L. Yu, H. Xu, H. Zhang, T. Nguyen, Vector sparse representation of color image using quaternion matrix analysis. IEEE Trans. Image Process. 24(4), 1315–1329 (2015)
119. R.E. Blahut, Fast Algorithms for Signal Processing (Cambridge University Press, Cambridge, 2010)
120. R.M.C. de Souza, H.M. de Oliveira, A.N. Kauffman, A.J.A. Paschoal, in Information Theory, 1998. Proceedings. 1998 IEEE International Symposium. Trigonometry in finite fields and a new Hartley transform (IEEE, 1998), p. 293
121. J.B. Lima, R.M.C. Souza, Finite field trigonometric transforms. Appl. Algebra Eng. Commun. Comput. 22(5–6), 393–411 (2011)
122. J.B. Lima, F. Madeiro, F.J.R. Sales, Encryption of medical images based on the cosine number transform. Signal Process. Image Commun. 35, 1–8 (2015)
123. J.B. Lima, L.F.G. Novaes, Image encryption based on the fractional Fourier transform over finite fields. Signal Process. 94, 521–530 (2014)
124. Ell, T.A.: Quaternion-Fourier transforms for analysis of two-dimensional linear time-invariant partial differential systems. In: Proceeding of the 32nd Conference on Decision and Control, San Antonio, TX, pp. 1830–1841 (1993)

125. Georgiev, S., Morais, J.: Bochner's theorems in the framework of quaternion analysis. In: Hitzer, E., Sangwine, S.J. (eds) Quaternion and Clifford Fourier Transforms and Wavelets. Trends in Mathematics. Birkhäuser, Basel (2013)
126. Georgiev, S., Morais, J., Kou, K.I., Sprößig, W.: Bochner-Minlos Theorem and Quaternion Fourier Transform. In: Hitzer, E., Sangwine, S.J. (eds) Quaternion and Clifford Fourier Transforms and Wavelets. Trends in Mathematics. Birkhäuser, Basel (2013)
127. Georgiev, S., Jday, R.: Bochner-Minlos theorem in the frame of real Clifford algebras. Rend. Circ. Mat. Palermo, II. Ser (2020). https://doi.org/10.1007/s12215-020-00487-5
128. El Haoui, Y., Fahlaoui, S.: Miyachi's Theorem for the Quaternion Fourier Transform. Circ. Syst. Sig. Process 39, 2193–2206 (2020). https://doi.org/10.1007/s00034-019-01243-6
129. Bahri, M.: Quaternion algebra-valued wavelet transform. Appl. Math. Sci. 5(71) 3531–3540 (2011)
130. Bahri, M., Lawi, A., Aris, N., Saleh, A.F., Nur, M.: Relationships between convolution and correlation for Fourier transform and quaternion Fourier transform. Int. J. Math. Anal. 7(43), 2101–2109 (2013)
131. Hitzer, E.: Directional uncertainty principle for quaternion Fourier transforms, Adv. Appl. Clifford Algebra 20(2), 271–284 (2010). https://doi.org/10.1007/s00006-009-0175-2, preprint: http://arxiv.org/abs/1306.1276
132. Abouelaz, A., Achak, A., Daher, R., Safouane, N.: Donoho–Stark's uncertainty principle for the quaternion Fourier transform. Bol. Soc. Mat. Mex. (2019). https://doi.org/10.1007/s40590-019-00251-5
133. Lian, P.: Sharp Hausdorff–Young inequalities for quaternion Fourier transforms. Proc. Am. Math. Soc. (2019). https://doi.org/10.1090/proc/14735
134. Loualid, E.M., Elgargati, A., Daher, R.: Quaternion Fourier transform and generalized Lipschitz Classes. Adv. Appl. Clifford Algebr. 31(14), 271–284 (2021). https://doi.org/10.1007/s00006-020-01098-0
135. Castro, L. P., M. R. Haque, M. M. Murshed, S. Saitoh, and N.M. Tuan. 2014. Quadratic Fourier transforms. Annals of Functional Analysis 5 (1): 10–23.
136. M.Y. Bhat, A.H. Dar, The algebra of 2D Gabor quaternionic offset linear canonical transform and uncertainty principles. J. Anal. (2021). https://doi.org/10.1007/s41478-021-00364-z
137. Bhat, M. Y., and A. H. Dar. 2021. Donoho-Stark's and Hardy's uncertainty principles for the short-time quaternion offset linear canonical transform. arXiv:2110.02754v1.
138. Guanlei, X., Xiaotong, W., Xiaogang, X.: Uncertainty inequalities for linear canonical transform. IET Signal Process. 3(5), 392–402 (2009). https://doi.org/10.1049/iet-spr.2008.0102
139. Zhang, Y.N., Li, B.Z.: Novel uncertainty principles for two-sided quaternion linear canonical transform. Adv. Appl. Clifford Algebras 28(1), 15 (2018)
140. Zhang, Y. N., and B. Z. Li. 2018. Generalized uncertainty principles for the two-sided quaternion linear canonical transform. In: International Conference on Acoustics, Speech and Signal Processing, pp. 15–20.
141. Shah, F.A., W.Z. Lone, and A.Y. Tantary. 2021. Short-time quadratic-phase Fourier transform. Optik - International Journal of Light Electron Optics. https://doi.org/10.1016/j.ijleo.2021.167689.
142. Shah, F.A., K.S. Nisar, W.Z. Lone, and A.Y. Tantary. 2021. Uncertainty principles for the quadratic-phase Fourier transforms. Mathematical Methods in the Applied Sciences. https://doi.org/10.1002/mma.7417.
143. Brackx, F., Delanghe, R., Sommen, F.: Clifford Analysis, Pitman Research Notes in Mathematics, vol. 76. Pitman, London (1982)
144. Delanghe, R., Sommen, F., Soucek, V.: Clifford Algebra and Spinor Valued Functions: A Function Theory for the Dirac Operator, vol. 53. Kluwer Academic, Dordrecht (1992)
145. Gilbert, J., Murray, M.: Clifford Algebras and Dirac Operators in Harmonic Analysis. Cambridge University Press, Cambridge (1991)
146. Sudbery E.: Quaternionic analysis. Math. Proc. Camb. Philos. Soc. 85(2), 199–225 (1979)
147. Rudin, W.: Function Theory in the Unit Ball of C^n . Springer, Berlin (1980)

148. Chen, L., Zhao, J.: Weyl transform and generalized spectrogram associated with quaternion Heisenberg group. Bull. Sci. Math. 136(2), 127–143 (2012)
149. Ghosh, S., Srivastava, R. K.: Unbounded Weyl transform on the Euclidean motion group and Heisenberg motion group, arXiv:2106.15704
150. Peng, L., Zhao, J.: Weyl transforms associated with the Heisenberg group. Bull. Sci. Math. 132(1), 78–86 (2008)
151. Peng, L., Zhao, J.: Weyl transforms on the upper half plane. Rev. Mat. Complut. 23(1), 77–95 (2010)
152. Gröchenig, K., Jaming, P., Malinnikova, E.: Zeros of the Wigner distribution and the short-time Fourier transform. Rev. Mat. Complut. 33(3), 723–744 (2020)
153. Parui, S., Thangavelu, S.: On theorems of Beurling and Hardy for certain step two nilpotent groups. Integral Transforms Spec. Funct. 20(1–2), 127–145 (2009)
154. Helgason, S.: The Radon Transform, Progress in Mathematics, 5, Birkhäuser. Mass, Boston (1980)
155. Ghiloni R, Perotti A. Slice regular functions on real alternative algebras. Adv Math, 2011, 226(2): 1662–1691
156. Sitaram, A., Sundari, M., Thangavelu, S.: Uncertainty principles on certain Lie groups. Proc. Math. Sci. 105, 135–151 (1995)
157. Faress, M., Fahlaoui, S.: Spherical Fourier transform on the quaternionic Heisenberg group. Integr. Transform Spec. Funct. 31(9), 685–701 (2020)
158. Fei, M., Xu, Y., Yan, J.: Real Paley–Wiener theorem for the quaternion Fourier transform. Complex Var. Elliptic Equ. 62, 1072–1080 (2017)
159. Yang, Y., Dang, P., Qian, T.: Tighter uncertainty principles based on quaternion Fourier transform. Adv. Appl. Clifford Algebras 26, 479–497 (2016)
160. B. Mawardi, A modified uncertainty principle for two-sided quaternion Fourier transform. Adv. Appl. Clifford Algebras 26(2), 513–527 (2016)

Quaternionic Moments

Eckhard Hitzer

Chapter Introduction This short chapter surveys five papers containing a variety of quaternionic moment applications in fields like color image watermarking, image representation and recognition, bio-signal watermarking, and quaternionic fractional-order pseudo-Jacobi Fourier moments.

Machine Generated Keywords moment, copyright, watermarking, image, orthogonal, watermark, radial, technology, invariant, zero, recognition, host image, protect, descriptor, polar

1 Robust Color Image Watermarking Using Multi-core Raspberry Pi Cluster

DOI: https://doi.org/10.1007/s11042-022-12037-5

Abstract-Summary
Watermarking is one of the many ways used to protect transmitted images.

Digital image watermarking technology is used to secure and ensure digital images' copyright by embedding hidden information that proves its copyright.

The color images Parallel Robust watermarking algorithm using the Quaternion Legendre-Fourier Moment (QLFM) in polar coordinates is implemented on the Raspberry Pi (RPi) platform with parallel computing and C++ programming language.

E. Hitzer (✉)
International Christian University, Mitaka, Tokyo, Japan
e-mail: hitzer@icu.ac.jp

E. Hitzer (ed.), *Quaternionic Integral Transforms*, Trends in Mathematics,
https://doi.org/10.1007/978-3-031-28375-8_4

The watermarking algorithm is implemented and tested on a Raspberry Pi model 4B. We can combine many Raspberry Pi's into a 'cluster' (many computers working together as one) for high-performance computation.

The Message Passing Interface (MPI) and OpenMP for parallel programming accelerate the execution time for the color image watermarking algorithm implemented on the Raspberry Pi cluster.

Extended:

Digital image watermarking technology secures sensitive digital images by embedding hidden information that proves these images' integrity [1].

Introduction
Moment-based watermarking algorithms are successfully used in robust color image watermarking due to the quaternion moments' exactness, speed, and computational stability.

Much research work has been performed using the credit-card-sized open-source Raspberry Pi platform in various fields such as image processing [2, 3, 4], the internet of things (IoT) [5, 6, 7], home automation [8, 9], and many other applications.

Many research works have been done using the Raspberry Pi platform in various fields such as image processing, due to its portability, and because it can be controlled via the internet and its low cost [10].

The contributions of this paper are summarized as: A new parallel QLFM watermarking algorithm for color images is implemented on Raspberry Pi.

The portable and small-sized embedded system (Raspberry Pi) is used to overcome the limited portability of the previous watermarking algorithms executed on a P.C. The Message Passing Interface (MPI) and a four-nodes Raspberry Pi cluster are used to accelerate the watermarking processing to be suitable for real-time applications.

QLFM Color Image Watermarking
The input host image of size N × N in Cartesian coordinate is first converted to the polar domain using cubic interpolation [11].

After computing QLFM moments of the host color image, we select suitable QLFM moments for embedding the watermark that achieve these two factors: QLFM moments with q = 4 m are ignored. Only the positive repetition q is selected to avoid information redundancy.

Raspberry Pi Implementation
The Message Passing Interface (MPI) is a library for writing parallel programs to accelerate the computing time using a cluster compared to a single computer; a cluster consists of at least two nodes.

The Message Passing Interface (MPI) is an interface that allows the head node (Master node) to distribute the computing task among all the slave nodes in the cluster in a parallel way as needed.

There is a continually increasing demand for small, portable, and low-cost, powerful computers that provide powerful computational power, increasing the computational power can be done by parallel computing, parallel computing on a Raspberry

Pi cluster can achieve the computational power needs in applications where time is a critical metric in these applications, besides the low cost of the Raspberry Pi cluster that makes it a better choice.

A cluster consists of at least 2 nodes to facilitate parallel computing with distributed memory, and the Message Passing Interface (MPI) is required.

Experiment Results and Discussion

A parallel robust QLFM color image watermarking algorithm using a C++ programming language is implemented on a Raspberry Pi model 4 cluster, the cluster consists of 4 Raspberry Pi's, the RPi's cluster degrades the time needed for the color watermarking algorithm, several experiments were done on a different number of RPi's cluster cores and different image sizes and with moment order 40.

Applying the BER function on our algorithm shows a similar watermark image extraction as the original binary watermark image embedded without extraction errors.

The authors performed experiments on different cluster cores with moment order 40 and color image size 512 × 512 and 256 × 256 and a binary watermark image of size 32 × 32.

To measure the efficiency of the watermarking algorithm implementation on the Raspberry Pi cluster, we calculate the speedup and the execution time improvement ratio.

Conclusions

The transmission of sensitive images over the internet is still an issue, and a challenge against attackers—the limited portability of the P.C. limits smart-cities for new applications to be applied.

Searching for outstanding performance, lower power consumption, portability, and lower cost simultaneously, we can use a cluster of Raspberry Pi's to perform the watermarking task in a few seconds.

For smart-cities watermarking application needs and portability, this technique uses the C++ programming language and parallel computing on a Raspberry Pi's cluster that shows excellent performance comparing an ordinary expensive P.C. with low Raspberry Pi specifications with its low cost and small size.

Acknowledgement

A machine generated summary based on the work of Hosny, Khalid M.; Magdi, Amal; Lashin, Nabil A.; El-Komy, Osama; Salah, Ahmad

2022 in Multimedia Tools and Applications

2 Helmet-Fourier Orthogonal Moments for Image Representation and Recognition

DOI: https://doi.org/10.1007/s11227-022-04414-6

Abstract-Summary
The orthogonal moments have recently achieved outstanding predictive performance and become an indispensable tool in a wide range of imaging and pattern recognition applications, including image reconstruction, image classification and object detection.

Using these functions we introduce three new sets of orthogonal moments and their invariants to scaling, rotation and translation for image representation and recognition, named, respectively, "the orthogonal helmet-Fourier moments" for gray-level images, the multi-channel orthogonal helmet-Fourier moments and the quaternion orthogonal helmet-Fourier moments (QHFMs) for color images.

The performance of these feature vectors is compared with existing orthogonal invariant moments.

Extended:

Orthogonal moments presented in polar coordinates are often defined on a circular domain such as Zernike moments [12, 13], substituted and weighted radial shifted Legendre moments [14], pseudo-Zernike moments [15], Chebyshev-Fourier moments [16], Fourier-Mellin moments [17], radial harmonic Fourier moments [18], Bessel-Fourier moments [19], polar harmonic transforms [20], exponent-Fourier moments [21] and polar harmonic Fourier moments [22] defined in polar coordinates over a disk of radius 1.

We will extend and apply the proposed p-orthogonal moments to other pattern recognition tasks, such as facial expression recognition [23], pose-invariant face recognition [24] and Human Emotion Detection [25].

Introduction
Invariant moments have been used in many image analysis and pattern recognition applications in recent years, including visual image recognition [26], image storage and retrieval [27], target recognition of objects in infrared images [28], analysis of English and Chinese letters and characters [29], pedestrian detection using saliency maps [30], action recognition from point cloud patches [31], image denoising [32], face recognition [33], color image representation [34], color image texture analysis [35], 3D image recognition [36], content-based image retrieval [37], image analysis [38], robust copy-move forgery-detection [39], pattern classification [40], 2D and 3D object recognition through sketches [41], scene matching [42], etc. Extracting characteristics using non-orthogonal invariant moments was first introduced by Hu in 1962 [26].

The modulus of these moments is invariant under rotation, which is a very important property in image recognition.

Based on polar coordinates, the descriptor vectors for color images can be expressed in quaternion orthogonal or multi-channel invariant moments as quaternion polar harmonic Fourier moments for color images [43], quaternion orthogonal Zernike moments (QZMs) presented by Chen and others [44], multi-channel orthogonal radial substituted Chebyshef moments (MRSCMs) proposed by Hosny and others [45] and the multi-channel orthogonal Zernike moments (MZMs) presented by Singh and others [46], the quaternion orthogonal Zernike moments

(QZMs) [46], the quaternion pseudo-Zernike moments (QPZMs) [47], the quaternion Fourier-Melin moments (QFMMs) [48], the quaternion orthogonal generalized Chebyshef-Fourier (QGCFMs) and pseudo-Jacobi Fourier moments (QPJFMs) [49].

Three new sets of orthogonal moments and their invariants to scaling, rotation and translation (SRT) for image representation and recognition: are the orthogonal helmet-Fourier moments (HFMs) for grayscale images, the multi-channel orthogonal helmet-Fourier moments (MHFMs) and the quaternion orthogonal helmet-Fourier moments (QHFMs) for color images.

The orthogonal helmet-Fourier moments (HFMs) and their invariants for gray-level images are presented in the third section.

Orthogonal Helmet Functions

In the third part, we define the set of helmet-Fourier orthogonal functions in polar coordinates.

Multi-channel Orthogonal Helmet-Fourier Moments for Color Images

This section contains two subsections; the multi-channel orthogonal helmet-Fourier moments (MHFMs) for color images are presented in the first subsection.

The rotation, translation and scaling invariants of the (MHFMs) are shown and derived in the second subsection.

Quaternion Helmet-Fourier Moments for Color Images

In this section, we present the orthogonal quaternion helmet-Fourier moments (QHFMs) for color images, and in the second one, we derive invariants under rotation, translation and scaling (QHFMs).

By eliminating the effect of the rotation we can obtain new invariants of quaternion orthogonal helmet-Fourier moments (NQHFMs). To ensure the translation invariance of QHFMs we transform the origin of coordinates to the common centroid of the input RGB color image.

The Adopted Computational Methods

To perform the calculation of the previous moments, we adopt the rectangle-to-circle technique.

This technique is presented by Xin and others and improved by Hosny and others [50, 51].

Experimental Analysis

Presented in these tables, we can say that the invariant moments (MHFMs) and (QHFMs) have better performance than the other tested invariant moments for the three affine image transformations.

In this sub-section, we will provide experiments to validate the precision of recognition and the classification of gray-level images, according to the object, using the proposed HFM orthogonal invariant moments.

We performed our classification system and the previous comparative study on the COIL-20 database under various conditions: rotated, scaled and translated dataset, and we have concluded that our proposed orthogonal invariant moments (HFMs) are robust for the three geometric transformations despite the noisy conditions and the recognition criterion works better than the other descriptors tested.

To evaluate the performance of the proposed orthogonal moments MHFMs and QHFMs under STR transformations and their robustness to noise, we present five experiments performed on the last four color image databases under four different conditions: rotation, translation, scale, noise and normal.

Conclusion

We have proposed three types of orthogonal moments for image representation and recognition: the orthogonal helmet-Fourier moments (HFMs) for gray-level images, the multi-channel orthogonal helmet-Fourier moments (MHFMs) and the quaternion orthogonal helmet-Fourier moments (QHFMs) for color images.

A comparative performance analysis has been made between the proposed orthogonal invariant moments and existing orthogonal moments on the recognition of objects and in image classification using the Euclidean distance.

The proposed orthogonal invariant moments perform better in image recognition tasks.

The image reconstruction by the proposed orthogonal moments is performed by a growth of the clear circular rings according to the maximum order.

The proposed orthogonal moments are more invariant under the three geometric transformations of translation, scaling and rotation.

In future work, we will extend and apply the proposed p-orthogonal moments to other pattern recognition tasks, such as facial expression recognition [23], pose-invariant face recognition [24] and Human Emotions Detection [25].

Acknowledgement

A machine generated summary based on the work of Hjouji, Amal; EL-Mekkaoui, Jaouad

2022 in The Journal of Supercomputing

3 Fast Quaternion Log-polar Radial Harmonic Fourier Moments for Color Image Zero-watermarking

DOI: https://doi.org/10.1007/s10851-022-01084-0

Abstract-Summary

Three issues of the moments-based zero-watermarking methods should be addressed: first, most of them ignore the analysis and experiment on discriminability, which leads to a high false positive ratio; second, direct computation of the moments from their definition is inefficient, numerically unstable and inaccurate, which severely affects the performances of these moments-based methods; third, most of the algorithms are only suitable for grayscale images, which leads to a limitation of the algorithms in practical applications.

We present a new color image zero-watermarking using fast quaternion log-polar radial harmonic Fourier moments.

This algorithm firstly introduces log-polar coordinates to make traditional RHFMs scale-invariant, secondly improves their speed and accuracy, then introduces the concept of quaternion to make it suitable for color images, and finally applies it to the zero-watermark technology.

Extended:

The discrimination is verified by experiments.

Introduction

For decades, some zero-watermarking schemes have been developed for image copyright protection.

According to important image features, the existing zero-watermarking approaches can be roughly classified into three groups: (1) spatial domain features based, (2) frequency domain features based, and (3) moments-based methods.

To overcome the aforementioned critical issues, in this paper, we present a new color image zero-watermarking using fast quaternion log-polar radial harmonic Fourier moments (FQLPRHFMs).

We develop a new color image zero-watermarking approach using FQLPRHFMs and asymmetric tent mapping.

Related Work

Compared with the zero-watermarking algorithm of spatial features, the zero-watermarking algorithm based on the frequency domain still has high robustness against the conventional signal attack, but because it does not possess rotation invariance, the expressive force is poor at the time of resisting the geometric attack, and the time complexity is high.

Kang and others [52] proposed a robust and distinguishable color image zero-watermarking algorithm based on polar harmonic transforms (PHTs) and compound chaotic maps.

Wang and others [53] proposed two image zero-watermarking methods based on moments, one is based on Quaternion Exponential Moments (QEMs), and the other is zero-watermarking methods based on the Polar Harmonic Transform (PHT) and logistic mapping.

Based on ternary number theory and radial harmonic Fourier moments (RHFM), Wang and others [54] introduced ternary radial harmonic Fourier moments (TRHFMs) to deal with stereo images in a holistic manner, and then proposed a robust stereo image zero-watermarking algorithm using TRFFM.

Radial Harmonic Fourier Moments

Ren and others [55] suggested to choose a triangular function as the radial function and introduced new orthogonal moments named radial harmonic Fourier moments (RHFMs).

Compared to other orthogonal moments, RHFMs have a better image reconstruction, lower noise sensitivity, and the property of rotation invariance, but they lack natively the scaling-invariant property.

This drawback influence seriously the practical application of RHFMs.

Fast Quaternion Log-Polar Radial Harmonic Fourier Moments (FQLPRHFMs)
We introduce a new framework for computing the RHFMs in the log-polar domain by using a fast pseudo-polar Fourier transform and frequency domain interpolation, which results in better image representation capability, numerical stability, and computational speed.

We extend the RHFMs' basis functions from the Cartesian/polar coordinate domain to the log-polar domain, so that the proposed moments have an inherent property of scaling and rotation invariance.

To extend nice properties of the proposed FLPRHFMs to color image processing, we generalize FLPRHFMs from the complex field to the hypercomplex field, and introduce fast quaternion log-polar radial harmonic Fourier moments (FQLPRHFMs) based on quaternion algebra.

We will first demonstrate the intrinsic relationship between the quaternion log-polar RHFMs and log-polar RHFMs, and then use the FLPRHFMs to achieve fast calculation of QLPRHFM coefficients for color digital images.

We believe that the FLPRHFMs algorithm can be used for the implementation of quaternion log-polar RHFMs for color digital images.

Color Image Zero-Watermarking Based on FQLPRHFMs
The zero-watermark generation mainly discusses how the copyright owner uses the carrier image to generate an effective zero-watermark, and send the zero-watermark and key together to a trusted institution for verification.

The zero-watermark verification mainly discusses how the copyright verifier uses the carrier image to be verified, zero-watermark and key to detect the image copyright information.

The copyright owner is responsible for using the zero-watermarking generation algorithm to construct copyright verification information; the copyright verifier may be anyone who wants to verify the copyright of the carrier image; the TA is mainly responsible for authentication and ensuring the integrity and authenticity of the relevant data used by the owner and verifier.

Zero-watermark verification does not require the original carrier image, and only requires the verifier to have relevant information such as zero-watermark and key.

Experimental Results
The experiment on the zero-watermarking scheme mainly includes five aspects, namely, robustness, discriminability, watermark capacity, speed and security.

We mainly test the robustness of the proposed zero-watermarking approach for common image processing operations and geometric attacks.

We further evaluate the performance of the proposed zero-watermarking and compare it to the latest method proposed in [56, 57, 53, 58].

Four medical images are selected from the "whole brain atlas" to compute the BER values to verify the robustness of the proposed zero-watermarking algorithm.

The proposed zero-watermark method has better discriminability.

From experimental results in terms of PSNR, BER, FPR, AR and time obtained over standard color images and color medical images and different size digital watermarks, it is evident that the proposed color image zero-watermarking approach achieves high work performance compared with other methods.

We proposed a new color image zero-watermarking approach using FQLPRHFMs and asymmetric tent map.

The novel contributions of this work can be summarized as follows. (1) We extended the RHFMs' basis functions from the Cartesian/polar coordinate domain to the log-polar domain, and obtained the quaternion log-polar RHFMs with inherent scaling and rotation invariance properties. (2) We derive the fast quaternion log-polar RHFMs (FQLPRHFMs) by using a fast pseudo-polar Fourier transform and frequency domain interpolation and quaternion algebra, which results in better image representation capability, numerical stability, and computational speed. (3) We developed a new color image zero-watermarking approach using FQLPRHFMs and the security of the algorithm is increased by using the asymmetric tent map.

The proposed method cannot effectively deal with image cropping; hence, future work will investigate possible combinations of the FQLPRHFMs descriptor with local descriptors.

Acknowledgement
A machine generated summary based on the work of Niu, Pan-Pan; Wang, Li; Wang, Fei; Yang, Hong-Ying; Wang, Xiang-Yang
2022 in Journal of Mathematical Imaging and Vision

4 New Method for Bio-signals Zero-watermarking Using Quaternion Shmaliy Moments and Short-time Fourier Transform

DOI: https://doi.org/10.1007/s11042-022-12660-2

Abstract-Summary
This descriptor is applied to the copyright protection of bio-signals after converting the latter into color spectrogram images using the Short-Time Fourier Transform (STFT).

The proposed method for bio-signal copyright protection is implemented via a novel zero-watermarking scheme.

The simulation and comparison results prove on the one hand the numerical stability of the proposed computation of high-order SPs, and on the other hand, they demonstrate the robustness of the proposed zero-watermarking scheme against various signal-processing attacks (compression, filtering, noise, etc).

Extended:

The proposed method consists in using the Short Term Fourier Transformation (STFT) to convert the bio-signal into a spectrogram that will be stored as a color image.

The proposed method interrupts the recursive computation of SPs when the polynomial values become inferior to a low threshold ($T \leq 10^{-6}$), thus stopping the propagation of the numerical errors that lead to the appearance of unstable values.

A conclusion and future work are listed in the last section.

In the future, we will focus on other applications of QSMs such as pattern recognition and classification of color images.

Introduction

Only a few studies have been conducted in the area of 1D signal copyright protection based on zero watermarking schemes.

Ali and others [59] have proposed a zero-watermarking algorithm for privacy protection of medical speech signals.

A new method for biomedical signal zero-watermarking is proposed in the present work.

The main differences between our work and that of Ali and others are that (i) we use a new color image descriptor that is QSMs to extract the features of the time-frequency spectrogram after converting the latter into a color image via STFT, and (ii) we implement the proposed method via a novel bio-signal zero-watermarking scheme.

The proposed zero-watermarking scheme for the copyright protection of bio-signals is presented in the fourth section.

Discrete Orthogonal Shmaliy Polynomials Computation

The following Algorithm 2 summarizes the proposed recursive computation of high order SPs according to the order n. Algorithm 2: Algorithm in pseudo-code for the proposed computation of higher-order SPs.

Algorithm 3: Algorithm in pseudo-code for higher-order SPs computation using GSOP.

This problem led us to look for an alternative method to GSOP that guarantees the numerical stability of high order SPs with minimal computational cost.

We first analyze the numerical behavior of SPs values in the polynomial matrix that is computed by the GSOP method.

These values are still very low and tend towards zero, which confirms that our supposition is true, and it is valid to ensure the numerical stability of SPs for any order n with low computational cost.

After solving the problems that limit the computation of high order SPs, we can now use these polynomials to compute Shmaliy moments (SMs) of large-size signals/images.

Shmaliy Moments and Quaternion Shmaliy Moments

The following transformation is used to obtain the reconstructed 2D image from SMs of order (n, m): In this section, we present a new type of color image descriptor called Quaternion Shmaliy Moments (QSMs).

The following relation defines the module of q: The use of pure quaternions makes it possible to represent an f(x, y) color image [60]: where $f_R(x, y)$, $f_G(x, y)$ and $f_B(x, y)$ represent the red, green and blue channels of the color image, respectively.

The color image representation using pure quaternions allows characterization of the whole of this image instead of characterizing each channel separately.

Proposed Scheme for Signal Zero-watermarking Based on QSMs and STFT

The values of ε, η and γ are given as security key denoted as KEY 2, and the permuted logo image is noted as $L_p = \{l_p(x, y), 1 < x, y \leq n\}$. Zero watermark image generation to create the zero-watermark (ZW) image, the "or-exclusive" operation is applied for L_p and C_p as follows: The five steps are applied to k non-overlapping frames of the original bio-signal to generate a k zero-watermark image.

For each frame S∗, the corresponding spectrogram image is generated, and the feature vector V∗ is constructed based on the method described in section 4.1. Construction of the feature image: The characteristic image C∗ is obtained from V∗ by using the same method described in the Section 4.2 in the feature image permutation In this step, C∗ and B matrices are used to create the permuted feature image C_p∗. For recovering the binary logo image, the "or-exclusive" operation is used to reproduce the permuted logo image, where ZW is the zero-watermark image and C_p∗ represents the permuted feature image.

Simulation Results

We present the simulation results and comparisons that demonstrate the numerical stability of high-order SPs computed by the proposed method (Algorithm 2), and validate the robustness of the proposed bio-signal zero-watermarking scheme.

To validate the efficiency of the proposed zero-watermarking scheme for the copyright protection of bio-signals, we test its robustness against various common signal-processing attacks (noise, filtering, scaling, cropping, compression, etc).

From the obtained results, it can be noticed that the BER value corresponding to each recovered watermark tends toward zero, which is a clear sign of the robustness of the proposed method against signal-shifting attack.

The low BER values achieved by the proposed method is explained by the fact that our method is implemented in the time-frequency domain, where the robustness against signal processing attacks is better compared to implementation in the time domain.

Conclusion

Based on STFT and QSMs, a new robust scheme for bio-signal zero - watermarking has been proposed.

The simulation results have proven the efficiency of the proposed computation of high order SPs.

These results have shown the robustness of the proposed zero-watermarking scheme against different types of signal processing attacks.

Acknowledgement

A machine generated summary based on the work of Daoui, Achraf; Karmouni, Hicham; Sayyouri, Mhamed; Qjidaa, Hassan

2022 in Multimedia Tools and Applications

5 Accurate Quaternion Fractional-order Pseudo-Jacobi–Fourier Moments

DOI: https://doi.org/10.1007/s10044-022-01071-6

Abstract-Summary
Pseudo-Jacobi–Fourier moments (PJFMs) are a set of orthogonal moments which have been successfully applied in the fields of image processing and pattern recognition.

We present a new set of quaternion fractional-order orthogonal moments for color images, named accurate quaternion fractional-order pseudo-Jacobi–Fourier moments (AQFPJFMs).

We define a new set of orthogonal moments, named accurate fractional-order pseudo-Jacobi–Fourier moments, which is characterized by its generic nature and time–frequency analysis capability.

We finally extend the gray-level fractional-order PJFMs to color images and present the quaternion fractional-order pseudo-Jacobi–Fourier moments.

Introduction
As early as 1962, Hu [26] proposed using geometric moments to describe images.

This method mapped the entire square image to the unit disk and calculated integrals separately like precise geometric moments.

Compared with traditional orthogonal moments that can only reflect global information, fractional orthogonal moments can better describe the image information of the region of interest.

It is also applied to the new radial basis functions to define fractional orthogonal moments.

To make the fractional orthogonal moments not only limited to grayscale images, Chen [39] proposed fractional-order orthogonal quaternion Zernike moments (FQZMs) to calculate color images.

In order to design an ideal moment descriptor, we propose a new algorithm called accurate quaternion fractional-order Pseudo-Jacobi–Fourier moments (AQFPJFMs) in this paper.

The polar pixel tiling scheme can cancel geometric and numerical integration errors, so we can reconstruct high-order images better.

Accurate Pseudo-Jacobi–Fourier Moments
We can get the original image through the above mentioned division method to obtain the pixel division in polar coordinates.

We use the pixel sampling method to solve this question, then the gray value of the new pixel can be estimated through the points in the original image.

The new complete orthogonal basis functions can be defined. Studies have found that recursive methods [61] to calculate high-order polynomials can also produce numerical instability.

This method can maintain numerical stability when calculating the integral of high-order polynomials.

Accurate Fractional-order Pseudo-Jacobi–Fourier Moments
Unlike classical PJFMs, the fractional-order parameter can control the AFPJFMs' zeros distribution, which shows that the focus described by AFPJFMs can be transferred to a higher resolution image area.

The moment value describes the entire image area equally.

The moment value emphasizes the information in the central area of the image.

The moment value emphasizes the information in the outer area of the image.

Accurate Quaternion Fractional-order Pseudo-Jacobi–Fourier Moments
In this section, we will extend AFPJFMs to the color image domain and propose accurate quaternion fractional-order Pseudo-Jacobi–Fourier moments (AQFPJFMs) to suit color image processing.

Theoretically, AQFPJFMs are not scaling invariance.

In the calculation of an AQFPJFMs' process, the image functions are all normalized and mapped to the unit circle.

We can consider the AQFPJFMs coefficients to be scaling invariance.

Image Representation via Mixed Low-order Moments
Low-order moments reflect more robust low-frequency information, global feature descriptions usually use low-order moments to improve robustness.

Features only containing a few low-order moments usually cannot distinguish multiple objects in an image.

We propose a strategy called mixed low-order moments feature (MLMF).

Experiments have proved that AQFPJFMs hybrid low-order moment features have good robustness.

Experiments and Application
Moments based on the zero-watermarking algorithm still have problems to be solved, including: The traditional zero-watermarking algorithm is generally applicable to the protection of grayscale images, in reality, color images to be protected are more common; the traditional zero-watermarking algorithm paid too much attention to robustness, ignoring the distinction.

This phenomenon leads to a high false detection rate; most zero-watermarking algorithms are based on frequency domain features, their features are not robust to geometric attacks.

This is because the AQFPJFMs' low-order moments and MLMF algorithms are more robust to conventional image attacks (sharpening, blurring, compression, noise, etc).

Since most zero-watermarking algorithms are publicly verifiable, in addition to their robustness, discriminability is also extremely important.

It can be seen that compared with other watermarking algorithm methods, our proposed zero-watermarking algorithm performs better in both robustness and discriminability.

Conclusion

First, we proposed the recursive method of PJFMs, because the recursive calculation method does not involve the factorial (or gamma function) of large numbers, so it has better numerical stability; second, we proposed an accurate algorithm of PJFMs, called APJFMs.

It has been improved in terms of accuracy, numerical stability and applicability; finally, we make full use of the time–frequency analysis capabilities of AQFPJFMs to propose the MLMF and apply it to the digital image zero-watermarking algorithm.

Compared with other advanced zero-watermarking algorithms, our proposed algorithm proves that the MLMF has significant advantages in both global feature robustness and discriminability.

According to testing, our proposed zero-watermarking algorithm has high security and has certain advantages in zero-watermarking capacity and calculation speed.

Acknowledgement

A machine generated summary based on the work of Wang, Xiangyang; Zhang, Yuyang; Tian, Jialin; Niu, Panpan; Yang, Hongying

2022 in Pattern Analysis and Applications

References

1. Begum M, Uddin MS (2020) Digital image watermarking techniques: a review. Information 11(2):110
2. Manikandan LC, Selvakumar RK, Nair SAH, Kumar KS (2020) Hardware implementation of fast bilateral filter and canny edge detector using Raspberry Pi for telemedicine applications. Journal of Ambient Intelligence and Humanized Computing 12(5):4689–4695
3. Sajjad M, Nasir M, Muhammad K, Khan S, Jan Z, Sangaiah AK, Baik SW (2020) Raspberry Pi assisted face recognition framework for enhanced law-enforcement services in smart cities. Futur Gener Comput Syst 108:995–1007
4. Widodo CE, Adi K, Gunadi I (2020) The use of raspberry pi as a portable medical image processing. In Journal of Physics: Conference Series, vol 1524, no 1. IOP Publishing, Bristol, p 012004
5. Ahmad I, Pothuganti K (2020) Design & implementation of real-time autonomous car by using image processing & IoT. In 2020 Third International Conference on Smart Systems and Inventive Technology (ICSSIT). IEEE, pp 107–113
6. Atmaja AP, Hakim E, Wibowo A, Pratama LA (2020) Communication systems of smart agriculture based on wireless sensor networks in IoT. J Robot Control (JRC) 2(4):297–301
7. Petrović N, Kocić Đ (2020)IoT-based system for COVID-19 indoor safety monitoring. IcETRAN Belgrade
8. Akour M, Radaideh A, Shadaideh K, Okour O (2020) Mobile voice recognition based for smart home automation control. Int J Adv Trends Comput Sci Eng 9(3):3788–3792. https://doi.org/10.30534/ijatcse/2020/196932020
9. Iromini NA, Alimi TA (2020) Development of a Raspberry Pi secured management system for home automation. i-Manager's. J Embed Syst 8(2):1
10. Sagar S, Choudhary U, Dwivedi R (2020) Smart home automation using IoT and Raspberry Pi. Available at SSRN 3568411

11. Xin Y, Pawlak M, Liao S (2007) Accurate computation of Zernike moments in polar coordinates. IEEE Trans Image Process 16(2):581–587
12. Kim WY, Kim YS (2000) A region-based shape descriptor using Zernike moments. Signal Process: Image Commun 16:95–102
13. Kanaya N, Liguni Y, Maeda H (2002) 2-D DOA estimation method using Zernike moments. Signal Process 82:521–526
14. Xiao B, Wang G, Li W (2014) Radial shifted legendre moments for image analysis and invariant image recognition. Image Vis Comput 32:994–1006
15. Bailey R, Srinath M (1996) Orthogonal moment features for use with parametric and non-parametric classifiers. IEEE Trans Pattern Anal Mach Intell 18:389–399
16. Ping ZL, Wu R, Sheng YL (2002) Image description with Chebyshev-Fourier moments. J Opt Soc Am A 19:1748–1754
17. Sheng Y, Shen L (1994) Orthogonal Fourier-Mellin moments for invariant pattern recognition. J Opt Soc Am A 11:1748–1757
18. Ren H, Ping Z, Bo W, Wu W, Sheng Y (2003) Multidistortion-invariant image recognition with radial harmonic Fourier moments. J Opt Soc Am A 20:631–637
19. Xiao B, Ma J, Wang X (2010) Image analysis by Bessel-Fourier moments. Pattern Recognit 43:2620–2629
20. Yap P-T, Jiang X, Kot AC (2010) Two-dimensional polar harmonic transforms for invariant image representation. IEEE Trans PAMI 32(6):1259–1270
21. Hu H-T, Zhang Y-D, Shao C, Ju Q (2014) Orthogonal moments based on exponent functions: exponent-Fourier moments. Pattern Recognit 47:2596–2606
22. Wang C, Wang X, Xia Z, Ma B, Shi Y-Q (2019) Image description with polar harmonic fourier moments. IEEE Trans Circuits Syst Video Technol 30(12):4440–52
23. Naveen P, Sivakumar P (2021) A deep convolution neural network for facial expression recognition. J Current Sci Technol 11(3):402–410
24. Naveen P, Sivakumar P (2021) Adaptive morphological and bilateral filtering with ensemble convolutional neural network for pose-invariant face recognition. J Ambient Intell Human Comput 12:10023–10033. https://doi.org/10.1007/s12652-020-02753-x
25. Naveen P, Sivakumar P (2021) Human emotions detection using kernel nonlinear collaborative discriminant regression classifier : human emotions detection using KNCDRC. In: 2021 2nd International Conference on Smart Electronics and Communication (ICOSEC), 1807–1812, doi: https://doi.org/10.1109/ICOSEC51865.2021.9591878
26. Hu M-K (1962) Visual pattern recognition by moment invariants. Inf Theory IRE Trans On 8:179–187
27. Teague MR (1980) Image analysis via the general theory of moments. J Opt Soc Am 70:920–930
28. Zhang F, Liu SQ, Wang DB, Guan W (2009) Aircraft recognition in infrared image using wavelet moment invariants. Image Vis Comput 27:313–318
29. Hjouji A, Bouikhalene B, EL-Mekkaoui J et al (2021) New set of adapted Gegenbauer Chebyshev invariant moments for image recognition and classification. J Supercomput 77:5637–5667
30. Lahouli I, Karakasis E, Haelterman R, Chtourou Z, Cubber GD, Gasteratos A, Attia R (2018) Hot spot method for pedestrian detection using saliency maps, discrete Chebyshev moments and support vector machine. In: IET Image processing, Vol. 12, pp 1284–1291
31. Hjouji A, Chakid R, El-Mekkaoui J et al (2021) Adapted jacobi orthogonal invariant moments for image representation and recognition. Circuits Syst Signal Process 40:2855–2882
32. Ji Z, Chen Q, Sun Q-S, Xia D-S (2009) A moment-based nonlocal-means algorithm for image denoising. Inf Process Lett 109:1238–1244
33. Hjouji A, El-Mekkaoui J, Qjidaa H (2021) New set of non-separable 2D and 3D invariant moments for image representation and recognition. Multimed Tools Appl 80:12309–12333

34. Hosny KM, Darwish MM (2018) New set of quaternion moments for color images representation and recognition. J Math Imaging Vision 60:717–736
35. Assefa D, Mansinha L, Tiampo KF, Rasmussen H, Abdella K (2010) Local quaternion Fourier transform and color image texture analysis. Signal Process 90:1825–1835
36. Batioua I, Benouini R, Zenkouar K, Zahia A, Hakim EF (2017) 3D image analysis by separable discrete orthogonal moments based on Krawtchouk and Tchebichef polynomials. Pattern Recognit 71:264–277
37. Singh C, Pooja (2012) Local and global features based image retrieval system using orthogonal radial Moments. Opt Lasers Eng 50:655–667
38. Xiao B, Li L, Li Y, Li W, Wang G (2017) Image analysis by fractional-order orthogonal moments. Inf Sci 382–383:135–149
39. Chen B, Yu M, Su Q, Shim HJ, Shi YQ (2018) Fractional quaternion Zernike moments for robust color image copy-move forgery detection. IEEE Access 6:56637–56646
40. Hmimid A, Sayyouri M, Qjidaa H (2015) Fast computation of separable two-dimensional discrete invariant moments for image classification. Pattern Recognit 48:509–521
41. Ansary TF, Daoudi M, Vandeborre J-P (2007) A Bayesian 3D search engine using adaptive views clustering. IEEE Trans Multimed 9:78–88
42. Lin YH, Chen CH (2008) Template matching using the parametric template vector with translation, rotation and scale invariance. Pattern Recognit 41:2413–2421
43. Wanga C, Wang X, Li Y, Xiac Z, Zhang C (2018) Quaternion polar harmonic Fourier moments for color images. Inf Sci 450:141–156
44. Chen BJ, Shu HZ, Zhang H, Chen G, Luo LM (2012) Quaternion Zernike moments and their invariants for color image analysis and object recognition. Signal Process 92:308–318
45. Hosny KM, Darwish MM (2019) New set of multi-channel orthogonal moments for color image representation and recognition. Pattern Recognit 88:153–173
46. Singh C, Singh J (2018) Multi-channel versus quaternion orthogonal rotation invariant moments for color image representation. Digital Signal Processing 78:376–392
47. Chen BJ, Sun XM, Wang DC, Zhao XP (2012) Color face recognition using quaternion representation of color image. ACTA Automatica Sinica 8:1815–1823
48. Guo L, Zhu M (2011) Quaternion Fourier-Mellin moments for color images. Pattern Recogn 44:187–195
49. Singh C, Singh J (2018) Quaternion generalized Chebyshev-Fourier and pseudo Jacobi-Fourier moments. Opt Laser Technol 106:234–250
50. Xin Y, Pawlak M, Liao S (2007) Accurate computation of Zernike moments in polar coordinates. IEEE Trans Image Process 16:581–587
51. Hosny KM, Shouman MA, Abdel Salam HM (2011) Fast computation of orthogonal Fourier-Mellin moments in polar coordinates. J Real-Time Image Proc 6:73–80
52. Kang, X., Zhao, F., Chen, Y., Lin, G., Jing, C.: Combining polar harmonic transforms and 2D compound chaotic map for distinguishable and robust color image zero-watermarking algorithm. J. Vis. Commun. Image Represent. 70, 102804 (2020)
53. Wang, C.P., Wang, X.Y., Xia, Z.Q., Zhang, C.: Geometrically resilient color image zero-watermarking algorithm based on quaternion Exponents. J. Vis. Commun. Image Represent. 41, 247–259 (2016)
54. Wang, C., Wang, X., Xia, Z., Zhang, C.: Ternary radial harmonic Fourier moments based robust stereo image zero-watermarking algorithm. Inf. Sci. 470, 109–120 (2019)
55. Ren, H., Ping, Z., Bo, W., Wu, W.: Multidistortion-invariant image recognition with radial harmonic Fourier moments. J. Opt. Soc. Am. A 20(4), 631–637 (2003)
56. Sun, L., Xu, J.C., Zhang, X.X.: A novel generalized Arnold transform-based zero-watermarking scheme. Appl. Math. Inf. Sci. 4, 2023–2035 (2015)
57. Guo, Y., Liu, C.P., Gong, S.R.: Improved algorithm for Zernike moments. In: International Conference on Control, Automation and Information Sciences (ICCAIS), pp. 307–312 (2015)

58. H. Y. Yang, S. R. Qi, P. P. Niu, X. Y. Wang. Color image zero-watermarking based on fast quaternion generic polar complex exponential transform. Signal Processing: Image Communication, 2020, 82:115747.
59. Ali Z, Imran M, Alsulaiman M, Zia T, Shoaib M (2018) A zero-watermarking algorithm for privacy protection in biomedical signals. Future Gener Comput Syst 82:290–303
60. Sangwine SJ, Ell TA (2001) Hypercomplex Fourier transforms of color images. In: Proc. 2001 Int. Conf. Image Process. Cat No 01CH37205. IEEE, pp 137–140
61. Walia E, Singh C, Goyal A (2012) On the fast computation of orthogonal Fourier–Mellin moments with improved numerical stability. J Real-Time Image Process 7(4):247–256

Octonion Fourier Transform

Eckhard Hitzer

Chapter Introduction This short chapter surveys five publications that go beyond the generalization of complex Fourier transforms to quaternions by introducing the non-associative highest dimensional normed division algebra of eight dimensions: octonions. Quaternions are a natural subalgebra of octonions, in fact the Caley Dickson construction shows how pairs of quaternions can be combined to octonions. Our survey shows applications to the analysis of multidimensional linear time-invariant systems, to three-dimensional linear time-invariant differential systems, and to real valued Lipschitz functions of three variables. Finally, an octonion version of offset linear canonical transforms is introduced.

Machine Generated Keywords Octonion, OFT, octonion fouri, Snopek, realvalue, hypercomplex, ell, transform oft, previous, variable, realvalue function, Hahn Snopek, Hahn, theory, smoothness

1 Hypercomplex Fourier Transforms in the Analysis of Multidimensional Linear Time-Invariant Systems

DOI: https://doi.org/10.1007/978-3-030-27550-1_73

Abstract-Summary
The aim of this paper is to further investigate the properties of the octonion Fourier transform (OFT) of real-valued functions of three variables and its potential applications in signal and system processing.

E. Hitzer (✉)
International Christian University, Mitaka, Tokyo, Japan
e-mail: hitzer@icu.ac.jp

E. Hitzer (ed.), *Quaternionic Integral Transforms*, Trends in Mathematics,
https://doi.org/10.1007/978-3-031-28375-8_5

This is a continuation of the work started by Hahn and Snopek, in which they studied the octonion Fourier transform definition and its applications in the analysis of hypercomplex analytic signals.

The main part of the article is devoted to new properties of the OFT, that allow us to use the OFT in the analysis of multidimensional signals and LTI systems, i.e. derivation and convolution of real-valued signals.

Introduction

Hypercomplex Fourier transforms deserve special attention in this consideration.

The quaternion Fourier transform (QFT) allows us to analyze two dimensions of the sampling grid independently, while the complex transform mixes those two dimensions.

In [1] we presented some preliminary results concerning the octonion Fourier transform (OFT).

There are known results for the QFT (see [2]), but they use the notion of another hypercomplex algebra, i.e. double-complex numbers.

Octonion Fourier Transform

The octonion Fourier transform (OFT) of u is defined where $x = (x_1, x_2, x_3)$, $f = (f_1, f_2, f_3)$ and multiplication is done from left to right.

Conditions of existence (and invertibility) are the same as for the classical (complex) Fourier transform.

In [1] we derived basic properties of the OFT, analogous to the properties of the classical Fourier transform.

We have the octonion analogue of Hermitian symmetry. Moreover, if U^{α}, U^{β} and U^{γ} denote the OFTs of functions $u(x_1 - \alpha, x_2, x_3)$, $u(x_1, x_2 - \beta, x_3)$ and $u(x_1, x_2, x_3 - \gamma)$, respectively, then we have the octonion version of the shift theorem.

Recent Results

The next result concerns function convolution.

The convolution-multiplication duality is one of the key properties used in the frequency analysis of LTI systems [2].

As in the previous theorem, this result follows from expressing the OFT as a sum of components of different parity.

Multidimensional Linear Time-Invariant Systems

It is a well-known fact that the Fourier transform converts differential equations into algebraic equations.

We loose the information that the function u was differentiated at all.

Further information indicates that the function has been differentiated by x_1 and x_2.

Final Remarks

The presented results further develop the foundation of octonion-based signal and system theory.

We are left to find real-life applications of the discussed theory.

It seems that extending octonion-based signal theory to discrete-variable signals may also be achieved by methods used so far.

Acknowledgement
A machine generated summary based on the work of Błaszczyk, Łukasz 2019 in Mathematics in Industry.

2 Discrete Octonion Fourier Transform and the Analysis of Discrete 3-D Data

DOI: https://doi.org/10.1007/s40314-020-01373-7

Abstract-Summary
The purpose of this article is to further develop the theory of octonion Fourier transformations (OFT), but from a different perspective than before.

The research described in this article applies to discrete transformations, i.e. discrete-space octonion Fourier transform (DSOFT) and discrete octonion Fourier transform (DOFT).

The described results combine the theory of the Fourier transform with the analysis of solutions for difference equations, using for this purpose previous research on algebra of quadruple-complex numbers.

Introduction
As in the case of classical signal processing, so the discrete counterpart of this theory has so far mainly focused on signals with real and complex values, as well as their complex spectra.

However, more and more works have started to appear, in which authors use in their research on hypercomplex algebras, among others quaternions and octonions (Brackx et al. [3]; Hahn and Snopek [4]; Lian [5]; Wang et al. [6]).

We focused on discussing the properties of the octonion Fourier transform (OFT) of real-valued functions of three variables (and in the case of some properties, we extended this to octonion-valued functions) (Błaszczyk [[7–9]; Błaszczyk and Snopek [1]]).

Non-commutativity is also encountered with Fourier transforms based on quaternion algebra and (in general) Clifford algebras (Brackx et al. [3]).

In the case of octonions, references to the analysis of 1- and 2-dimensional signals appear in the literature (Grigoryan and Agaian [10]), but there is still no definition of the discrete octonion Fourier transform of 3-dimensional signals.

Octonion Fourier Transform and Some Properties
In general, conditions of existence of the OFT are the same as for the classical (complex) Fourier transform (Błaszczyk [9]).

We have also proved some important features, among which there are those that will be useful in the analysis of discrete-variable signals (Błaszczyk [9]; Błaszczyk and Snopek [1]).

It seems that in the above form these statements are of little use.

It allows the direct use of the OFT for the analysis of LTI systems, which are described both by partial differential equations and difference equations (of three variables).

Discrete-Time LTI Systems

Consider a linear time-invariant stationary system of three variables.

Analogous reasoning can be performed for discrete-time systems.

Discrete-Space Octonion Fourier Transform

The above definition is a three-dimensional equivalent of the Fourier transformation of discrete time (discrete-time Fourier transform – DTFT), in relation to which the given formula can be abbreviated as DSOFT (discrete-space octonion Fourier transform).

Using the methods presented in the proof of the inverse theorem (Błaszczyk and Snopek [1]), the following formula can be derived for the inverse transform.

From simple calculations the proof of the following theorem on the transformation of the rescaled function follows.

Differences in statements of those theorems resulted only from the properties of multiplication of octonions, and, as a consequence, of operations on exponential functions in the kernel of the transformation.

The theory considered so far mainly used classic Fourier transforms (discrete), which were applied to a variable interpreted as space (one- or two-dimensional), but by defining an octonion transformation we can try to transform the whole scheme, both for time and space.

Discrete Octonion Fourier Transform

In practice, finite-length signals are usually found, which, as in the classical case, leads to the definition of a discrete octonion Fourier transform.

In the quaternion case, various solutions to this problem have been proposed – direct implementation of the quaternion version of the Fast Fourier Transform algorithm (FFT) and the use of the complex (original) version of this algorithm.

In Błaszczyk [9] we proved that the octonion Fourier transformation of octonion-valued functions can be calculated using the classical Fourier transformation.

Thanks to this, it is possible to use all the advantages of the FFT algorithm, with a small additional calculation effort – the octonion FFT algorithm for functions with octonion values requires the calculation of four transforms of different functions with complex values.

It uses the convention that MATLAB® adopts with the FFT algorithm – zero frequencies always appear on the first coordinates of the matrix (in each dimension), and the discrete Fourier transform is a periodic function.

Symmetry Properties of the DOFT

As in the case the of continuous OFT, most discrete classical equivalents of Fourier transform properties can be proved for the discrete OFT.

One of the first results for the continuous OFT was to show the equivalent of Hermitian symmetry (Błaszczyk and Snopek [1], Theorem 4.6).

In the discrete case, the following theorem can be proved by repeating the reasoning presented in Błaszczyk and Snopek [1].

Discussion and Conclusions
The results presented show that discrete Fourier transforms can be generalized to the case of higher order algebras (e.g. octonions).

It seems important to develop tools enabling work in this algebra as well.

The properties of the discrete octonion Fourier transforms show that they can be used without difficulty for the analysis of difference equations, as well as for the analysis of finite difference schemes for differential equations.

Acknowledgement
A machine generated summary based on the work of Błaszczyk, Łukasz 2020 in Computational and Applied Mathematics

3 A Generalization of the Octonion Fourier Transform to 3-D Octonion-Valued Signals: Properties and Possible Applications to 3-D LTI Partial Differential Systems

DOI: https://doi.org/10.1007/s11045-020-00706-3

Abstract-Summary
It is also a continuation and generalization of earlier work by Błaszczyk and Snopek, where they proved a few essential properties of the OFT of real-valued functions, e.g. symmetry properties.

The results of this article focus on proving that the OFT is well-defined for octonion-valued functions and almost all well-known properties of the classical (complex) Fourier transform (e.g. argument scaling, modulation and shift theorems) have their direct equivalents in the octonion setup.

Those theorems, illustrated with some examples, lead to the generalization of another result presented in earlier work, i.e. Parseval and Plancherel Theorems, important from the signal and system processing point of view.

There are known results for the QFT, but they use the notion of other hypercomplex algebra, i.e. double-complex numbers.

Extended:

Preliminary studies show that the notion of quadruple-complex numbers can be applied to define the DSOFT and to analyze linear difference equations.

Introduction
Fourier series and Fourier transforms enable us to look at the concept of a signal in a dual manner – by studying its properties in the time domain (or in the space domain in case of images), where it is represented by amplitudes of the samples (or pixels), or by investigating it in the frequency domain, where the signal can be

represented by infinite sums of complex harmonic functions, each with different frequency and amplitude (Allen and Mills [11]).

In some practical applications, signals are represented by more abstract structures, e.g. hypercomplex algebras (Ell et al. [12]; Grigoryan and Agaian [13]; Hahn and Snopek [4]; Snopek [14]).

They are defined on the basis of the Cayley–Dickson algebras and called the Cayley–Dickson Fourier transforms.

Analysis of the current state of knowledge on applications of octonions in signal processing shows some areas previously unexplored or requiring thorough theoretical and experimental studies, although some gaps have recently been filled (Błaszczyk and Snopek [1]; Błaszczyk [7, 8]; Lian [5]).

Basic Definitions

The definition of the octonion Fourier transform (OFT) of a real-valued function of three variables was introduced in Snopek [17] and used in later publications concerning theory of hypercomplex analytic functions (Hahn and Snopek [16]; Snopek [14, 15, 17, 18]).

In (Błaszczyk [7]) we stated that the inverse transform formula is correct for the octonion-valued functions and we presented the sketch of the proof.

It should be noted that in case of the octonion Fourier transform multi-channel signals (or, mathematically speaking, vector-valued functions) are treated and processed as an algebraic whole, without losing information about the relationship between individual channels (i.e. individual vector coordinates).

For many real- or complex-valued functions the form of the classic Fourier transform is well known.

We will start, as in the case of forward transform, from the case when a function has an OFT with complex values, but in the specific subfield of the octonion algebra it is enough to note that in every subfield of this type we can define the classic Fourier transform.

Properties of the Octonion Fourier Transform

One can notice that the cosine modulation theorem (with the cosine function as a carrier) is exactly the same as in the case of the complex Fourier transform.

In case of real-valued functions we already stated this theorem in our earlier work (Błaszczyk and Snopek [1]).

At the end of this section, we will cite several other results that are important from the point of view of system analysis, i.e. octonion analogues of Parseval-Plancherel Theorems for real-valued functions, which we proved in (Błaszczyk and Snopek [1]).

The above theorem shows that the OFT preserves the energy of octonion-valued functions.

It is worth mentioning the recent result in (Lian [5]), where the author argues that the OFT of an octonion-valued function also satisfies the Hausdorff-Young inequality.

Multidimensional Linear Time-Invariant Systems

We will focus on using the OFT and the notion of quadruple-complex numbers in the analysis of 3-D linear time-invariant (LTI) systems of linear partial differential equations (PDEs) with constant coefficients.

The classical Fourier transform is a well recognized tool in solving linear PDEs with constant coefficients due to the fact, that it reduces differential equations into algebraic equations (Allen and Mills [11]).

Both classic and octonion Fourier transforms of u are real-valued functions.

Additional theoretical considerations regarding partial differential equations and the use of integral transforms in Cayley–Dickson algebras can be found in (Ludkovsky [19]).

The notion of quadruple-complex number multiplication can be used to describe general linear time-invariant systems of three variables and to reduce parallel, cascade and feedback connections of linear systems into simple algebraic equations, as in classical system theory.

Discussion and Conclusions

It has been shown that the theory of octonion Fourier transforms can be generalized to the case of functions with values in higher-order algebras.

It can still be argued that, from a practical point of view, there are no visible advantages of using hypercomplex versions of Fourier transforms.

It remains to study the discrete case, i.e. discrete-space octonion Fourier transform (DSOFT).

Acknowledgement

A machine generated summary based on the work of Błaszczyk, Łukasz 2020 in Multidimensional Systems and Signal Processing

4 Octonion Fourier Transform of Lipschitz Real-Valued Functions of Three Variables on the Octonion Algebra

DOI: https://doi.org/10.1007/s11868-021-00405-y

Abstract-Summary

We use the octonion Fourier transform (OFT) of real-valued functions of three variables to prove the equivalence between K-functionals and the modulus of smoothness in the space of square-integrable functions (in Lebesgue sense).

Introduction

Researchers have been interested in Cayley–Dickson Fourier transforms which is a nontrivial generalization of the real and complex Fourier transform (FT) based on Cayley–Dickson algebras (see e.g. [20]).

The relation between smoothness conditions imposed on functions f(x) and the behavior of its Fourier transforms f near infinity is well known in the literature.

It is commonly known that studying the relation which exists between the smoothness properties of a function and the best approximations of this function in weight functional spaces is more convenient than usual with various generalized moduli of smoothness (see [21, 22]).

The K-functionals introduced by Peetre [23] are part of many problems of the theory of approximation of functions.

The examination of the relation which exists between the modulus of smoothness and K-functionals is known as one of the major problems in the theory of the approximation of functions.

Acknowledgement
A machine generated summary based on the work of Bouhlal, A.; Igbida, J.; Safouane, N.

2021 in Journal of Pseudo-Differential Operators and Applications

5 Octonion Offset Linear Canonical Transform

DOI: https://doi.org/10.1007/s13324-022-00705-6

Abstract-Summary
We introduce a novel integral transform namely octonion offset linear canonical transform (OCOLCT).

We first establish the fundamental properties associated with the octonion offset linear canonical transform (OCOLCT) and obtain a relationship between OCOLCT and the quaternion offset linear canonical transform (QOLCT).

Introduction
The hypercomplex Fourier transform has found applications in image filtering, color image processing, edge detection, watermarking, and pattern recognition [24–29].

The most important and basic hypercomplex Fourier transforms are the 2-D quaternion Fourier transforms (QFT).

In the modern signal processing community, the octonion Fourier transform (OFT) is becoming a hot trend for researchers.

The applications and the various properties of the OFT were established in [5, 9, 30].

W. B. Gao and B. Z. Li [31] extended the OFT to the linear canonical transform (LCT) domain, thus resulting in the formation of the octonion linear canonical transform (OCLCT).

They replaced the Fourier transform kernel by the LCT kernel and then established the various properties and the uncertainty principle for OCLCT.

We extend the OCLCT to a octonion offset linear canonical transform (OCOLCT) by replacing the LCT kernal by the offset linear canonical transform (OLCT) kernal.

Octonion Offset Linear Canonical Transform (OCOLCT)
Many known transforms are special cases of the OCOLCT.

The 1-D OCOLCT and the 1-D OFT of a signal f(x) are related, where the function h(x) is given (Inversion formula for one dimensional OCOLCT).

Properties of the OCOLCT
Before deriving some fundamental properties of the OCOLCT, we need to find the kernel expansion of the OCOLCT as follows.

Plancherel's theorem for the OCOLCT is given. The proof of the theorem is the same as for the OCLCT (Theorem 4; [31]), except the kernel of OCLCT is to be replaced by the OCOLCT kernel and Plancherel's theorem of the QLCT by Plancherel's theorem of the QOLCT.

Uncertainty Principles Associated with the OCOLCT
The uncertainty principle in harmonic analysis stems from the uncertainty principle in quantum mechanics.

In quantum mechanics the uncertainty principle states "more precisely the position of a particle is determined, the less precisely its momentum can be known and vice-versa".

In this section we will establish Hardy's uncertainty principle and the logarithmic uncertainty principle for the OCOLCT.

On the basis of the above proposition, we will derive the logarithmic uncertainty principle for the OCOLCT as follows.

Potential Applications of the OCOLCT
The results presented in this paper are theoretical in nature and can find applications in science and engineering, like image watermarking, signal processing, electrical and communication systems etc. By applying the OCOLCT, one can embed a color watermark into a selected part of the frequency modulated and time shifted octonion hyper-complex spectrum, this helps in analysing various watermarking techniques.

The host image (octonionic) by our proposed transform can be time shifted and frequency modulated and hence will improve the anti-attack capability of various watermarking algorithms.

The uncertainty principle defines that it is impossible to locate any signal sharply in both the time and frequency domain.

Hardy's uncertainty principle for the OCOLCT says that it is impossible for a non zero octonionic function and its OCOLCT to decrease simultaneously very rapidly.

Acknowledgement
A machine generated summary based on the work of Bhat, Younis Ahmad; Sheikh, N. A.

2022 in Analysis and Mathematical Physics

References

1. Błaszczyk, Ł., Snopek, K.M.: Octonion Fourier Transform of real-valued functions of three variables – selected properties and examples. Signal Process. 136, 29–37 (2017)
2. Ell, T.A.: Quaternion-Fourier transforms for analysis of 2-dimensional linear time-invariant partial-differential systems. In: Proceedings of 32nd IEEE Conference on Decision and Control, vols. 1–4, pp. 1830–1841 (1993)
3. Hitzer E, Sangwine S J. The Orthogonal 2D Planes Split of Quaternions and Steerable Quaternion Fourier Transformations//Hitzer E, Sangwine S. Quaternion and Clifford Fourier Transforms and Wavelets. Trends in Mathematics. Basel: Birkhüauser, 2013
4. Hahn SL, Snopek KM (2016) Complex and hypercomplex analytic signals: theory and applications. Artech House
5. Lian P (2019) The octonionic Fourier transform: Uncertainty relations and convolution. Sig Process 164:295–300. https://doi.org/10.1016/j.sigpro.2019.06.015
6. Wang R, Xiang G, Zhang F (2017) L1-norm minimization for octonion signals. In: 2016 International conference on audio, language and image processing (ICALIP), pp 552–556
7. Błaszczyk Ł (2018) Octonion spectrum of 3d octonion-valued signals—properties and possible applications. In: Proceedings of 2018 26th European signal processing conference (EUSIPCO), pp 509–513. https://doi.org/10.23919/EUSIPCO.2018.8553228
8. Błaszczyk Ł (2019) Hypercomplex Fourier transforms in the analysis of multidimensional linear time-invariant systems. In: Progress in industrial mathematics at ECMI 2018, pp. 575–581. Springer Nature Switzerland AG. https://doi.org/10.1007/978-3-030-27550-1_73
9. Błaszczyk Ł (2020) A generalization of the octonion Fourier transform to 3-d octonion-valued signals—properties and possible applications to 3-d LTI partial differential systems. Multidim Syst Sign Process 31(4):1227–1257. https://doi.org/10.1007/s11045-020-00706-3
10. Grigoryan AM, Agaian SS (2018) Quaternion and octonion color image processing with MATLAB. SPIE
11. Allen, R. L., & Mills, D. (2003). Signal analysis: time, frequency, scale, and structure. Hoboken: Wiley-IEEE Press.
12. Ell, T.A., Bihan, N.L., Sangwine, S.J.: Quaternion Fourier Transforms for Signal and Image Processing. Wiley, New York (2014)
13. Grigoryan, A.M., Agaian, S.S.: Quaternion and Octonion Color Image Processing with MATLAB. SPIE, Bellingham (2018)
14. Snopek, K. M. (2015). Quaternions and octonions in signal processing-fundamentals and some new results. Telecommunication Review + Telecommunication News, Tele-Radio-Electronic, Information Technology, 6, 618–622.
15. Snopek, K. M. (2009). New hypercomplex analytic signals and Fourier transforms in Cayley–Dickson algebras. Electronics and Telecommunications Quarterly, 55(3), 403–415.
16. Hahn, S. L., & Snopek, K. M. (2011). The unified theory of n-dimensional complex and hypercomplex analytic signals. Bulletin of the Polish Academy of Sciences: Technical Sciences, 59(2), 167–181.
17. Snopek, K.M. (2011). The n-d analytic signals and Fourier spectra in complex and hypercomplex domains. In Proceedings of 34th international conference on telecommunications and signal processing, Budapest, (pp. 423–427).
18. Snopek, K. M. (2012). The study of properties of n-d analytic signals in complex and hypercomplex domains. Radioengineering, 21(2), 29–36.
19. Ludkovsky, S. (2010). Analysis over Cayley–Dickson numbers and its applications. Saarbrucken: LAP LAMBERT Academic Publishing.
20. Hahn, S.L., Snopek, K.M.: The unified theory of n-dimensional complex and hypercomplex analytic signals. Bull. Polish Ac. Sci., Tech. Sci. 59(2), 167–181 (2011)
21. Potapov, M.K.: Application of the operator of generalized translation in approximation theory. Vestnik Moskovskogo Universiteta, Seriya Matematika, Mekhanika 3, 38–48 (1998)

22. Platonov, S.S.: An analogue of the Titchmarsh theorem for the Fourier transform on locally compact Vilenkin groups, p-Adic numbers. Ultrametric Anal. Appl. 9(4), 306–313 (2017)
23. Peetre, J.: A theory of interpolation of normed spaces, notes de Universidade de Brasilia, (1963)
24. C. J. Evans, S. J. Sangwine, T. A. Ell, Colour-sensitive edge detection using hypercomplex filters, 10th European Signal Processing Conference IEEE, (2000), 1–4
25. Gao, C., Zhou, J., Lang, F., Pu, Q., Liu, C.: Novel approach to edge detection of color image based on quaternion fractional directional differentiation. Adv. Autom. Robot. 1, 163–170 (2012)
26. S. C. Pei, J. J. Ding, J. Chang, Color pattern recognition by quaternion correlation, IEEE International Conference Image Process, Thessaloniki, Greece, October 7–10 (2010), 894–897
27. Sangwine, S.J., and T.A. Ell. 2000. Colour image filters based on hypercomplex convolution. IEE Proceedings-Vision, Image and Signal Processing 49 (21): 89–93.
28. Took, C.C., Mandic, D.P.: The quaternion LMS algorithm for adaptive filtering of hypercomplex processes. IEEE Transactions on Signal Processing 57(4), 1316–1327 (2009)
29. B. Witten, J. Shragge, Quaternion-based signal processing, Stanford exploration project, New Orleans Annu. Meet. (2006), 2862–2866
30. Blaszczyk, L.: Discrete octonion Fourier transform and the analysis of discrete 3-D data. Computational and Applied Mathematics 39(4), 1–19 (2020)
31. Gao, W.B., Li, B.Z.: The octonion linear canonical transform: definition and properties. Signal Processing (2021). https://doi.org/10.1016/j.sigpro.2021.108233

Printed in the United States
by Baker & Taylor Publisher Services